HE
8700.7
.C6
M175

Broadcasting and Cable

BROADCASTING AND CABLE

An Elementary Historical Programming Structure Sourcebook

M. Rogers "Boomer" McSpadden

VANTAGE PRESS
New York/Atlanta
Los Angeles/Chicago

FIRST EDITION

All rights reserved, including the right of
reproduction in whole or in part in any form.

Copyright © 1987 by M. Rogers McSpadden

Published by Vantage Press, Inc.
516 West 34th Street, New York, New York 10001

Manufactured in the United States of America
ISBN: 0-533-07171-2

Library of Congress Catalog Card No.: 86-90244

To Jean, Rogers, John, Tyler,
and the Tulsa broadcast media

Contents

Preface ix
Introduction xv

Chapter 1. Systems of Organization and Control 1
Chapter 2. Factors Affecting Types of Programs 12
Chapter 3. Radio and Television History—Development of Program Types 21
Chapter 4. The Broadcast Industry in the United States 66
Chapter 5. Classifications and Definitions 87
Chapter 6. Program Form Considerations 101
Chapter 7. Program Idea, Featured Personality, Program Formats 106
Chapter 8. Units of a Broadcast and Cable Program 117
Chapter 9. Requirements of Effective Program Structure 122
Chapter 10. The Purpose of a Broadcast or Cable Program 132
Chapter 11. Radio, Television, and Cable Ratings 135
Chapter 12. Variations in Program Ratings: The Reasons 146
Chapter 13. Appeals in Programs 153
Chapter 14. Appeals and Program Types 171
Chapter 15. Strength of Appeals for Different Types of Listeners and Viewers 181
Chapter 16. Potential Audiences and Coverage Areas 204
Chapter 17. Available Audiences and Selection of Broadcast Programs 225
Chapter 18. Audiences of Specific Programs 232

Chapter 19. Attention by Audiences to Programs
            Broadcast 246
Chapter 20. Advertising and Purpose Materials in Broadcast and
            Cable Programs 252
Chapter 21. Total Requirements of an Effective Program 262

Preface

BROADCASTING AND CABLE: AN ELEMENTARY HISTORICAL PROGRAMMING STRUCTURE SOURCEBOOK

This new book deals with program structure on radio, television, and cable in a historical perspective. It is a sourcebook with tests, or standards, that can be applied to programs to provide senior high school and undergraduate broadcast students with an understanding of program effectiveness. Since programs are broadcast with ideas of reaching and influencing people, my new book is also devoted to audiences reached by programs.

Today, more programs are available on radio, television, and cable than ever before, with an overwhelming majority of them commercial. The emphasis, therefore, is chiefly on commercial television and cable programming.

However, the same principles and tests of effectiveness discussed also apply to educational programs, video cassettes, or video disc offerings.

This textbook is primarily intended for the senior high school and freshman/sophomore college students. It should be used in introductory broadcast-journalism courses in senior high schools, broadcast-journalism sequences of course offerings in junior or community colleges, and as a secondary textual direction tool in programs and audiences courses in four-year journalism-broadcasting schools at universities.

Its aims are twofold: to develop an understanding of basic historical concepts and principles operating in the broadcast setting and to provide a sequential set of materials and structured experiences leading to improved understanding and performance by the student who will work in radio, television, and cable.

These two aims are interrelated and integrated as cogently and completely as possible. Throughout, the focus is upon the broadcast industry as a process—an ongoing, multidimensional phenomenon in which many variables are acting and interacting.

The major elements of the broadcast industry are: 1) agents, stations, networks, or cable outlets who 2) present and produce programs 3) for specific audiences in contexts. The situation whereby stations present programs for audiences suggests the archetype of all broadcast communication contexts; the principles are applicable and may be extended and generalized to other basic oral and visual communication settings.

Approximately one-half of my new book is devoted to the exposition of principles and the explication of constructs. The basic variables operating in the social importance of broadcasting; systems of organization and control; factors affecting overall programming; a capsulized history with graphs and charts showing program development periods; basic classifications and definitions; conventions regarding program form, ideas, and broadcast units, with analyses of those units; together with structural requirements, purposes of programs, program ratings, and variations in those ratings, are all surveyed on an elementary historical level.

The wide-ranging purposes served by these broadcast processes are examined, as are the effects produced by them. The historical development of distinct periods in programming history are traced from initial beginnings to implementation in decisive periods of development.

The other half of the book presents a simple sequence of exploratory, experiential, and instructional materials and study probes intended to assist the broadcast student improve his analytical and performative capabilities in actual broadcast communication situations. Based upon this central archetypal context, my book shows agents, stations, or cable outlets that present programs for audiences. This achieves maximally productive student feedback and interaction with classroom teachers and instructors.

The participatory/performative chapters progress from audience appeals, appeals relating to program types, program preferences of children and adults, as well as analysis of appeals in programs, and preferences for types of broadcast music, to television and cable program preferences.

Appearing at numerous junctures in the book are study probes in population analysis, appeals and listener types, potential audiences, analysis of audience response to appeals, station coverage available audiences, audiences of specific programs, analysis of audience response to appeals, selection of programs, attention given to programs, and total analysis of the effectiveness of programs.

These probes are closely interwoven within the fabric of the sourcebook, where the broadcast student's exposure to them will be most meaningful. Sometimes they serve to initiate and direct thinking about minority broadcasting or about concepts or problems in the broadcast industry; at other times, they exemplify or elaborate points such as social responsibilities of broadcasters, regulation of broadcasting, evaluating the American System, or broadcasting in the future. They are always designed to clarify, expand, and reinforce the broadcast student's understanding of essential ideas of key constructs.

The title of my textbook, *Broadcasting and Cable: An Elementary Historical Programming Structure Sourcebook*, contains a number of key terms that require students' immediate attention, if they, as students of broadcasting and cable communication, are to understand the primary objective of programming. However, after describing or, in some cases, hinting at the complex processes of developing and delivering successful programs, other books and works emphasize only two basic topical areas: (1) program preparation, and (2) program production.

These definitions, especially by Eastman and Klein, may be adequate for telling us what a program is and telling us how to go about organizing and presenting an "on-the-air" production in the most effective manner, but they fall short of the more thorough historical approach developed in my sourcebook.

First, others focus on specific elements in programming, such as program organization, treating each as a distinct element. Secondly, they often confine themselves to only two links, preparation and "live production," in the long chain of interacting events that actually make up the historical broadcasting or cable programming structural communication process.

I view programs and audiences as a processional structural inquiry, involving numerous elements that have historically been built upon each other. These elements reach back into the historical program development perspective and project into its aftermath. Rather than concentrating on specific functions as instructional entities, such as program preparation, presentation, organization, or style, my focus is on the interactive relationships historically and structurally among those behavioral elements that have had a profound effect on today's programming.

I define programs and audience structure as the product of behavioral inputs, actions, and outputs that are measured by audience behavior in relationship to programming objectives. My text uses a "system" approach that enables study of historical interaction and related elements as they directly affect the broadcast-cable industries as communication outcomes. I perceive programs and audiences as a process whose components are interrelated structurally; the historical study of programming shows structural inputs that produce defined outcomes in a given program audience.

I do not attempt to dictate the "correct" method or format for programs. I seek to provide the student with a sourcebook of informational tools needed for understanding and developing proper modes of understanding programs and audiences. Correctness is a situational phenomenon in current broadcast and cable programming. It is possible for networks, independent production companies, stations or cable entities to organize, structure, and deliver a program in typical textbook fashion, and fail to fulfill the desired purpose producers and advertisers set as initial goals. A program may stand out in the mind of a student as a "real jewel" with little thought for the goal and purposes in which the producer or advertiser wishes it to fit. What is proper in one situation, may, in fact, be incorrect in another. Thus, the successful student must be able to assess what broadcast programming structural events are needed to produce desired results in a specific program situation. This is what I want the student to accomplish.

A student will see an effective program as a successful problem-solver. The program must solve particular problems in the broadcast setting or process. And when the program solves those problems, the student will see the inputs that operate in the successful tests of effectiveness. Therefore, I have organized this text with a problem-solving and audience "system" mode in mind.

The information my book will give individuals is a remarkable historical "footing," in regard to programs and audiences, as well as a practical day-to-day insight for making applicable judgments and decisions in regard to ratings, program performance, and for potential, available, and specific program audiences in any specific market. These standards of effectiveness will give the student of broadcasting, lay person, or broadcast professional a set of structural standards that can be readily applied to individual programs.

As mentioned before, programs are broadcast with the idea of reaching and influencing people. The broadcast student, lay person, or broadcast professional, will readily see the effects of attention that must be given to programs and audiences in order to be effective. Also, these tests apply equally to educational or Public Broadcasting programs.

The work is primarily descriptive, with a small amount of quantitative work involved on an elementary level. These levels are those that professional broadcasters use every day.

M. Rogers "Boomer" McSpadden

Introduction

There is every reason to believe that radio and television, the broadcasting media, have a greater influence on opinions, attitudes, and behavior of the American people than any other single medium of communication or mass entertainment combined.

In support of this statement, attention is called to the following.

BROADCAST PROGRAMS ARE AVAILABLE IN EVERY AMERICAN HOME.

Since around 1950, between 97 and 98 percent of all homes have been equipped with radio receiving sets; as of the autumn of 1979, the best estimates are that all homes have radio sets, as well as automobiles.

The Television Advertising Bureau estimates that all homes have access to television programming, in addition to radio sets.

Throughout the nation, there are far more families with television sets than telephones, electric refrigeration, and bathtubs!

There are considerably more American homes that have television sets than receive a daily newspaper. With radio sets in every American home, and television sets in nearly all homes, broadcast programs are definitely available to all Americans.

PEOPLE SPEND HOURS EACH DAY LISTENING TO RADIO OR WATCHING TELEVISION PROGRAMS.

The A. C. Nielsen Research Company reported that during the twelve months ending in March 1959 in the average American television-equipped home, the family television set was in use for an average of 5.01 hours each day.

The figure was greater during winter/summer months—an average of 5.91 hours each day during January 1959—and lower during summer months—an average of 4.05 hours per day in August, 1958.

In the same television-equipped homes, a radio set was turned on for an average 1.75 hours each day, on a twelve-month-average basis.

This does not take into account radio listening in automobiles, offices, and places of business. This includes only the use of radio sets in the home.

Combined, either a television set or a radio set was in use in the average television-equipped home for a total of 6.75 hours a day, and this was thirty years ago!

Of course, the average individual in a family today does not listen to radio or watch television for that many hours each day. Some members of the family are out of the home during the hours when radio or television sets are in use.

But studies of individual listening and viewing suggest that the average woman spends from four to five hours a day listening to radio or watching television. The average man spends approximately three hours per day with the radio or television, and the average boy or girl of school age also spends three hours.

THE TIME SPENT LISTENING TO OR VIEWING BROADCAST PROGRAMS IS NEARLY TWICE THE AMOUNT OF TIME DEVOTED TO ALL OTHER AGENCIES OF MASS COMMUNICATION OR MASS ENTERTAINMENT COMBINED.

The average person—man, woman, boy, or girl—spends at least twenty hours a week with radio or television programs. This is a very conservative estimate.

How much time is given each week to other media of information or entertainment?

Men and women give thirty to forty-five minutes a day to newspaper reading and spend not more than four or five hours per week with the newspaper. Children, as a rule, spend less than half that time reading newspapers, including the comic sections.

Other reading might account for an additional two hours per week for adults, and not more than thirty minutes to an hour a week for children.

The average man or woman goes to the movies about once a month at most, with the figure higher for teenagers and subteen children. This represents no more than an hour a week for adults and children combined. Not more than an hour and a half a week is devoted to church attendance, attendance at public meetings, or going to the theater. Therefore, we can estimate that the time spent with sports events and all other types of entertainment amounts to no more than a half hour each week, for each person.

Combined, these various agencies of information or entertainment add up to not more than nine or ten hours each week, as compared with twenty hours of exposure to radio and television. Accordingly, we can say that radio and television certainly have the opportunity to influence the American people, if total time of exposure means anything.

THE LISTENER OR VIEWER PARTICIPATES IN A SITUATION, MAKING FOR A HIGH DEGREE OF SUGGESTIBILITY.

When the average person becomes a participant in the broadcast experience, he indulges in a form of entertainment. He's in his own home, seated in a comfortable chair, and relaxed. He is taking it easy, mentally as well as physically. He has no reason to be suspicious of suggestions made on programs. He has no reason to be critical or analytical.

Since the easiest thing to do is accept without question whatever is said, or whatever pattern of conduct suggested in the program, he accepts it. Some listening and viewing is done when the listener/viewer is not relaxed and seated comfortably, and when this is done to any great extent, it has traditionally involved the American housewife. She is usually sweeping, dusting, or washing dishes.

Physically, she isn't relaxed. Mentally, however, she is giving only partial attention to a program. Not critical of things heard or seen, she still, however, tends to accept ideas without much effort to evaluate their worth or merit.

Secondly, ideas presented in programs, or in commercials, are presented by people who are talking, as contrasted with ideas presented in impersonal "cold print" of a newpaper or magazine.

The listener/viewer knows the person who presents the ideas. That personality is part of the broadcast experience. Presumably, he likes that person and has some confidence in him, or he wouldn't be listening to or viewing the program.

Added to the tendency to listen uncritically is the personality of the radio or television entertainer who presents the ideas. This is a factor giving the listener/viewer an added reason to accept.

Thirdly, people have confidence in the integrity of radio and television. Numerous studies have been made relating believability of materials on radio or television to materials presented in printed form. Invariably, those who have been asked say they believe what is broadcast.

Significantly, 60 to 70 percent of those answering the question "If conflicting accounts of the same news event are printed in your daily newspaper and broadcast on your favorite radio or television station, which account would you believe?" believe that the "broadcast version" presents much more believability. The reason for this believability factor is the listener/viewer can see and hear, for himself, while not being forced to take someone else's word for what takes place.

In general, broadcasting networks and stations enjoy a high level of prestige with listeners and viewers. The listener/viewer believes the broadcaster is fair and unbiased in giving information, especially in comparison with newspapers, where news is affected by editorial policies of the owners or publishers.

In any event, people feel broadcast stations and networks are dependable and fair, adding to the likelihood that anything presented in a broadcast format will be accepted, more or less, uncritically by a large proportion of listening and viewing audiences.

These four facts account for the tremendous influence broadcasting exerts on ideas, opinions, beliefs, and behavior of listeners and viewers.

Two aspects of the total situation bear upon the total effectiveness of radio and television in modifying ideas. First, with so many stations and programs, the listener/viewer is likely to be exposed to several conflicting ideas. Consequently, he is left without fixed beliefs in the field involved.

If he hears six or eight programs, each advertising a different brand of tires, the total effect is not necessarily to "sell" him on any one brand, unless the commercials presented in behalf of one brand are more effective than commercials for competing brands.

Accordingly, if he is influenced by one musical program to a heightened appreciation for that style of music, he may be influenced in quite the opposite direction by the next musical program he selects.

Second, programs have "side influences," which are entirely unintentional as far as the broadcaster is concerned. For example, if the purpose of a given program is to sell tires and attract a given "target" audience through an adventure program shot on location in Hawaii, the picture of conditions in Hawaii may or may not be accurate. This, as well as commercials for tires, makes for an impression upon the listener/viewer.

Therefore, radio and television exercise a tremendous total influence on ideas, beliefs, attitudes, and behavior of listeners or viewers. It is difficult to know specifically the nature of the influence exerted, when one considers the multiplicity of programs encountered in the course of a month and "side influences" suggested.

It is, accordingly, the purpose of this book to explore programs and audiences, while developing some understanding of the American system of broadcasting and becoming aware of the influence it exerts upon the American public.

Broadcasting and Cable

Chapter 1

Systems of Organization and Control

Different countries have different forms and philosophies of government. The organization of broadcasting conforms to different philosophies of government.

Although the system of organization and control of radio, television, and cable is not exactly the same in any two major nations of the world, systems can be grouped generally under three major headings:

a) Ownership and operation of all broadcasting stations by the government, directly or through some semi-independent, government-owned corporation
b) Ownership and operation of all broadcasting stations by the private sector, with a minimum of government control
c) Ownership and operation through a mixed system, with stations owned and operated by the government or a government-owned corporation, and an equal number by private individuals

FIRST SYSTEM: GOVERNMENT OWNERSHIP

This is the system of organization found in practically all countries on the continent of Europe, Asia, and Africa. It conforms to the theory of government being all-powerful and controlled by either the public utilities or other agencies directly related to the public welfare.

Government-owned radio and television is in two forms, differing in the degree of control exercised by the government or by the political entity controlling the government.

Direct Government Ownership

This is the system of organization and control found in Russia, Poland, East Germany, and Iron Curtain countries. It is also used in Norway, Sweden, and Denmark. It can be found in a modified form in China, India, Pakistan, Southeast Asia, and the Middle East.

In this system, all radio, television, and cable stations, as well as networks, are owned directly by the government and operated by a branch of the government, usually called the Ministry of Education and Public Enlightenment.

Broadcasting is entirely noncommercial, with operations financed by an annual tax levied on all receiving sets. Throughout Europe, and other countries which have networks and cable facilities, virtually all programs are fed to stations, cable, and network facilities. They originate in a central location.

Russia's great size makes it an exception to this rule to some degree. Most programs are broadcast from several regional locations, without having a network connection, due to language barriers. With broadcasting directly under the control of the Central Committee of the Communist Party, nothing is broadcast running contrary to the interests of the Party. As programs, are put on the air with a strict propaganda objective, the Russian sphere of broadcasting expresses only a thorough political indoctrination. These programs are intended to spread political culture among listeners or viewers, particularly in the fields of music and drama. Nevertheless, other programs have a direct educational objective, with adult education programs relating to agriculture, homemaking, and certain programming provided for classroom use.

In direct-ownership countries, radio and television are not regarded as instruments of entertainment. Those who control broadcasting make little or no effort to give the listener or viewer what he wants. Instead, they give what those in power think the audience should have, in education, culture, and propaganda.

Control of a Government-Owned Corporation

This form of broadcasting originated in England in 1928, with the formation of the British Broadcasting Corporation. Great Britain uses this system of control for radio only. The special situation of television and cable will be considered later in this section.

The government-owned corporation is also the system of control for both radio and television in France, Italy, and the Union of South Africa. Under the government-owned corporation system, a corporation that is wholly owned by the government—similar to our public trusts or electric cooperatives—owns and operates all stations.

As in direct-government ownership countries, all stations are non-commercial, with operating costs paid from proceeds of a special tax levied each year on owners of receiving sets. Accordingly, programs originate in a central location, and are "fed" to stations throughout the country with virtually no local or regional origination.

The objective of broadcasting is primarily cultural and educational, with an option of giving the listener or viewer what those in control think the listener or viewer should have. Entertainment is provided, but it is secondary.

In Great Britain, the BBC operates all three radio networks, in a multiple network fashion. The "Home Service" is an easy-listening musical offering. The "Light Service," developed after World War II, provides primarily programs of an entertainment nature, while the third network service, known as the "Third Service," has a classical-music orientation.

France similarly provides service ranging from classical to light, on different networks, and a similar variation exists in Italy.

There is one outstanding difference between direct-government ownership and government-corporation systems. In the first, there is always the possibilitiy of government propaganda. In government-corporation operations, control is vested in a corporation independent of government and not under the control of the party in power.

In Great Britain, for example, no changes in personnel or policies of the BBC will take place when a conservative government is replaced by a labor government, or vice-versa.

At the same time, those who operate government-owned corporations are aware of the fact that the corporation, and consequently the entire structure of broadcasting, is supported by a tax, and that Parliament or legislative bodies can, if they so desire, abolish the tax. Consequently, the government-owned corporations do not unnecessarily antagonize the government. If a free discussion on radio or television takes place, based upon pending legislative issues which might prove embarrassing to the party in power, the issues generally aren't discussed on government-owned corporation radio, television, or cable facilities. Even in Great Britain, there is far less discussion of political, social, or economic issues than in the United States.

Television and Cable: The Special Case of Great Britain

Great Britain, originator of the government-corporation system, has been forced to switch to a different system in the case of television and cable.

British television, developed by the BBC, built one television station in each metropolitan area. But the revenue from the tax on TV receiving sets, in excess of $14.00 per year for each set-owning family, was not enough to permit establishment of a second television service. However, British television viewers demanded the opportunity to make a choice between competing programs, prior to cable hookups.

In 1955, consequently, the British Parliament created the Independent Television Authority (ITA) a government-owned corporation paralleling the BBC, though not deriving revenue from taxes on television receiving sets. The ITA, using borrowed money, built ITA-owned television stations in leading cities throughout Great Britain. It leased these ITA-owned stations on an individual rather than group basis to commercial contractors, with two or even three contractors dividing broadcast time on each station. For example, one contracting company would be given broadcasting rights Saturday and Sunday, while a second company would hold the rights during daytime hours on weekdays.

Similarly, the station in the next large metropolitan area would have a different set of contractors controlling the broadcast time. Each contractor provides programs for the broadcast hours authorized for contract time on the air, and sells "spot" announcements within, or between, programs to commercial advertisers.

As you can see, ITA stations, though government-corporation owned, are commercially operated. At the same time, the BBC provides noncommercial programs on another lineup of stations, which is owned and operated by the BBC on the same noncommercial basis as BBC radio. Accordingly, with the advent of cable programming, the same contractor relationship developed in that endeavor.

SECOND SYSTEM: PRIVATE OWNERSHIP AND OPERATION OF BROADCASTING AND CABLE

In this system of ownership and control, all stations are owned by private individuals, rather than by government agencies. There is no special tax on receiving sets. Costs of station operations are paid from revenue received from sale of time to advertisers. This system is in use in the United States, Spain, and in most of Latin America.

Use of private-ownership does not preclude ownership of some stations by the government, or government agencies. In the United States, however, there are a number of stations owned or operated by tax-supported state universities, local municipalities, and noneducational branches of state government, such as individual state departments of agriculture. These stations are operated as noncommercial, educational stations, with costs paid out of general tax money. In some Latin American countries, the government, through the Ministry of Education, may own four or five high-powered stations located in large cities. These stations are operated on a noncommercial basis.

Although the government does not own or operate stations in private ownership countries, it does have the responsibility of exercising a degree of control over the broadcast apparatus, through granting or withholding station licenses.

THIRD SYSTEM: MIXED SYSTEMS: GOVERNMENT AND PRIVATE OWNERSHIP

This is the system in use in three major countries: Canada, Japan, and Australia. In each, a government corporation has been created, modeled after the BBC, that owns and operates both radio and television stations. The government corporation in these countries is partly financed by a special annual tax levied on receiving sets. Government-corporation–owned radio stations are located in virtually every major city throughout these countries.

Side by side with the government-corporation entity are privately owned and operated commercial stations deriving all of their revenue from sale of time to advertisers. As commercial stations, with revenue from time sales, there are twice as many privately owned radio stations as there are stations owned by the government corporation.

The same is true in the case of television and cable. In each country, however, the first television stations were constructed by the government corporation, and although privately owned television stations were allowed in Japan at an earlier date, it was not until 1959 that the governments of Canada or Australia officially authorized construction of privately owned television stations and, eventually in the 1970s, allowed cable control in the private sector.

Programmming is a mixture of network and locally originating materials. In Canada and Australia, the only radio and television networks in existence are owned by the government corporation. With few exceptions, the Canadian Broadcasting Corporation's (CBC) radio stations are the high-powered, 50 kw stations, while the privately owned stations operate with more limited power. Privately owned stations sell time to local advertisers as well as "national spot" advertising to national advertisers. The CBC stations, including radio, television, and cable, do not sell time for local advertising.

However, CBC-owned networks carry commercial programming and feed these programs to affiliated stations, which include not only CBC-owned stations, but also a large proportion of privately owned stations. At the same time, the CBC provides a substantial number of cultural programs. These stations, and late cable facilities, also must carry a specified proportion of such programs to secure the CBC network commercial service.

The Canadian commercial stations were, until 1958, licensed by the CBC. An act of the Canadian Parliament, in 1958, created a new agency—the Canadian Board of Broadcast Governors—that was given the responsibility for licensing and regulating commercial stations.

In most respects, the system of broadcasting used in Australia parallels that in Canada, with high-powered government-corporation owned stations, and a considerably greater number of lower-powered, privately owned commercial stations. Two networks, both operated by the Australian Broadcasting Corporation (ABC), carry commercial programs to both ABC-owned and privately owned stations, with a strong emphasis on cultural programs, as well as commercial programming.

The American System of Broadcasting

On preceding pages, the systems of organization and control of broadcasting in other countries has been discussed in detail. It is best to discuss the American system of broadcasting, emphasizing organization and control, through a series of characteristics and effects.

Basic Characteristics

1. *Private ownership for stations.* In the U.S., the great majority of broadcasting stations are owned by private individuals or corporations, rather than by the government or agencies of the government. This is in line with basic concepts of free enterprise and freedom of the press.
2. *Wide diffusion of ownership.* No less important as a characteristic is the fact that station ownership is not concentrated in the hands of a few individuals or corporations, but is rather in the hands of a great number of different owning groups.

There are corporations and media groups that own more than one station, radio or television, and operate more than one cable facility, singly or in combination. There are over four thousand standard-band stations on the air, and of that number, only about one thousand are owned by owning groups with interests in more than one station, of approximately four hundred such owning groups represented.

In the case of television and cable, there is a somewhat greater degree of concentration; however, with a saturation level of approximately 4,000 stations on the air, 1500 are owned by approximately 500 multiple-station-owning groups, representing 400 to 500 owners.

To be sure, there is only a limited number of national networks, including cable network entities operating at the present time. But none of these own or control more than five VHF television stations or seven radio stations. Other stations served, retain full control over their own programming, rates, and business activities. Finally, the limited number of national networks is not a factor that seriously restricts wide diffusion of ownership of broadcast facilities.

3. *Commercial operations.* The third characteristic of American radio and television is that, aside from a limited number of state-owned educational stations, broadcasting in this country is commercial. Stations and networks alike, derive practically all of their revenue from sale of time to commercial advertisers.

4. *Competition.* The final characteristic of American radio and television centers upon the factor of competition. It is an intense climate of competition, with rivals in the community, service area, and for the same advertiser's dollar. They compete equally for attention of listeners and viewers. There is very low cooperation between any two networks, or stations in the same market. Every network and station is constantly working to earn a position of advantage over its competitors.

Effects of the American System of Broadcasting

Four basic characteristics of American broadcasting have a decided effect on the nature of programs provided by networks and stations. This is most strikingly illustrated if we compare American broadcasting with the type found in most European, government-ownership countries.

First, with more than three thousand different groups determining station policies, all competing actively with one another, plus several different networks competing with one another, there is no possibility of central planning of the entire system.

In a European country, all stations are owned and operated by a government agency, with programs planned and produced by that government agency. Therefore, broadcasting may be used to achieve any desired objective. There is centralized control, and that makes central planning of the entire system possible.

In America, however, with more than three thousand sets of stations owners all competing vigorously, the force of broadcasting is multidirectional. There is no effort to use programs to achieve any preselected social, economic, or political objectives. We don't find networks making any effort to avoid scheduling two quite similar programs at the same hour, to give viewers a greater choice. In the American system, there is a complete lack of centralized control.

A second result is that the major objective of broadcasters is providing entertainment. There is no more than a token effort to provide programs to educate, inform, raise standards of music appreciation, or raise cultural standards. Broadcasting is commercial.

While the advertiser wants to attract the largest possible audience for the program he puts on the air, entertainment is merely the means to get the largest possible audience. Only a limited number of viewers will watch educational, cultural, or socio-economic programs; therefore, advertisers wish to hold audiences, while providing programs with strong entertainment values.

Thirdly, it is the American audience that determines what programs will be broadcast and what will be kept off the air. Neither the advertiser nor broadcaster can afford to pay the costs for a program attractive to only the smallest number of viewers or listeners. If an audience likes a program, they will tune in that program, and consequently it gets a substantial number of needed listeners or viewers. If it is disliked, it will not be viewed and will have a very small audience.

Of course, programs with small audiences do not stay on the air. It is possible, however, in European countries, for those who control broadcasting to air types of programs those in power feel audiences should have. Therefore, with broadcasting supported by the government, broadcasters can afford to operate on such a basis. But, in this country, the disliked program will have too small an audience to warrant continuance. By tuning the set away from that channel, audiences themselves, rather than broadasters or advertisers, determine what is broadcast.

Fourthly, in this country, networks present a far greater number of elaborate and expensive programs than in Europe, or any country where the broadcast apparatus is government supported. The fact that radio and television are commercially supported means network and station revenue attain totals far higher than in Europe. The average American television network, for example, has revenues from sale of time amounting to considerably more than $400 million a year, compared with not more than $500,000 to $1 million per broadcast. This is in addition, to what the advertiser pays for the use of network time. There simply aren't any half-million or million-dollar programs in Europe, or even programs with production costs as high as $100,000. The commercial system in this country provides for the broadcasting of more expensive programs than would be possible in Europe. Intense competition between advertisers makes for great numbers of expensive programs each year. For example, there has been in recent years the development of mini-series, such as "The Winds of War" and "The Thornbirds," which are very expensive programs, on the television networks.

Another result is that in the United States, broadcasting techniques have developed much more rapidly than in government-ownership countries in Europe. This is the result of greater amounts of money available. More importantly, there is intense competition between networks. Each network wants to get ahead of competitors. Consequently, engineers are continually experimenting with new technical equipment that will add to the interest values of programs, including innovations in lighting, providing scenery, camera work, and color television.

Similarly, broadcasters who plan programs for networks are constantly experimenting with new program forms in the hope of attracting a greater number of viewers, whereas in Europe, attracting more viewers is not important. While the government-owned broadcasting agencies have no competition from other broadcasters, there is no need to develop new forms.

By way of illustration, when American occupation forces assumed control over broadcasting in the American zone in Germany and Austria after World War II, the only types of programs previously broadcast over stations in those countries were forms in use in the United States on network radio around 1929 and 1930. Audience participation programs, quiz programs, daytime serials, comedy drama, and interview programs were simply unknown in Europe.

The introduction of the ITA commercial television stations in Great Britain placed the BBC, for the first time, in a competitive situation in providing television programs.

The BBC was forced to modify its whole system of television programming, to meet competition of commercial stations and hold a share of the total audience. Competition between networks and stations, which characterizes American broadcasting, has certainly been an important factor leading to technical innovations and development of new programs.

Finally, in this country, far more programs dealing with controversial public issues are presented than in any European country. In Europe, with government broadcasting agencies depending on their government, rather than advertisers, for revenue, no broadcaster will air a program critical of the party in power. In most European countries, there is very little discussion of controversial issues, and discussion of only a very limited number in France and Great Britain.

In this country, on the other hand, there are no government restrictions to prevent discussion of controversial issues, and no dependence on the goodwill of the party controlling the legislature, since broadcasting derives none of its support from government appropriations.

Since many Americans are interested in vital national or local issues, broadcasters provide programs where such issues are discussed. In fact, as a result of pressure from the Federal Communications Commission, all three national television networks, as well as a number of cable facilities, are committed to one policy of devoting at least one hour or more each week in prime time to programs of a public interest nature. Practically all of these broadcasts are in the form of documentaries dealing with vital public issues, or two-sided discussions on such issues. The American system permits, and encourages, discussion of public issues in a degree unheard of in government-ownership countries in Europe.

Chapter 2

Factors Affecting Types of Programs

CHANGES IN PROGRAM OFFERINGS

Throughout the history of radio in the United States, there has been unceasing change in types of programs provided by networks and stations, and in the extent to which that program type is used. In the relatively short years of television, there has been a similar change in the nature of programs included in program schedules, especially network programming. By way of illustration, during the 1948–49 season, the four existing television networks devoted a total of fourteen hours a week of evening time to broadcasts of sports events. In 1949–50, fifteen evening hours a week of network time were used for sports broadcasts.

But, three years later, in 1952–53, networks scheduled only four to five evening hours a week for sports. More conspicuously, variety programs, including comedy-variety programs, were first scheduled on television networks. In 1948–49, nine hours a week were scheduled. In 1949–50, twelve hours a week were scheduled, and in 1950–51, twenty-two hours a week were scheduled.

But in 1951–52, the total had dropped to fourteen hours, and in 1953–54, ten hours. We have had seasons where networks have scheduled six to eight "amateur" or "talent contest" shows each week, only to see this form disappear entirely, later.

In some years, network schedules have included twenty-four half-hour detective-type programs. But a few years later, programs of this type had almost been discarded. In the late 1950s, we had a heavy upswing in adult western and jazz-detective thrillers. By the winter of 1959–60, both had passed the peak of popularity. In at least one season, networks scheduled ten to twelve musical variety programs, each built around an individual singing star, and the following year, only three or four survived.

REASONS FOR CHANGES IN TYPES OF PROGRAMS OFFERED

Changes in the kinds of programs made available to viewers, whether by network or single station, have not come about as a result of the whim of network executives, or by changing program preferences of network sponsors. The desires of network executives might be a highly important factor in a system of broadcasting in Europe; but in the U.S., there is no centralized control of networks and stations. They are very competitive for program offerings.

When changes take place, they come as the result of a combination of conditions and factors. Changes force broadcasters to offer programs of a certain type, or abandon formerly successful program types almost entirely. These factors change as years go by. However, changes in types of programs provided allows individual broadcasters little or no control.

Some factors will influence program types offered by networks and stations, simply by the genera of programs offered.

TECHNOLOGICAL IMPROVEMENTS IN BROADCASTING

Technological improvements in broadcasting and cable have caused immediate or later use of certain types of programs. For example, prior to 1925 and 1926, the only microphones in use had no more frequency response range than an ordinary telephone. This did not make for particularly effective broadcast music; however, some broadcast music was carried, despite microphone shortcomings.

But development of ribbon microphones, with much improved frequency characteristics and directional qualities, made broadcast music much more attractive. Again, until 1926, when "electronic pickup" of recorded music was introduced, "platter programs" on radio were not feasible. The development of recorded music was enhanced, then, in large part by the "electronic pickup," which made development of "platter shows" successful at a later date. This included a combination of new circumstances making the use of such programs desirable.

In the early days of television, when the iconoscope camera tubes were in use, the relative insensitivity of the iconoscope tube to light made outside pickups of sports events practically impossible, except on days when the sun shone with unusual brilliance. The development of the image orthicon tube, more sensitive and requiring much less light, resulted in greater use of outside sports events through broadcasting. It made studio programs much easier to present by reducing amounts of artificial lighting required, and consequently reducing the amount of heat beating down on entertainers "on camera."

The development of effective color television increased the use of spectacular special programs. A more recent development, over three decades ago, of videotaping made certain types of delayed broadcasts possible. These programs could not be put on the air at the time the event actually took place. Therefore, inventions and technological improvements have had a most profound effect on types of programs presented and types actually used.

INVENTION OF NEW PROGRAM FORMS

Not all inventions affecting broadcasting have been inventions on a technical or engineering scale. Originally, we had no innovative invention in the field of programming itself. In early days of broadcasting, prior to 1924, the types of programs provided fell principally into two categories: 1) talk shows, or 2) music programs, presented by vocal soloists or vocal duets, with very limited instrumental accompaniment, because studios were too small to permit use of full orchestras.

By 1928, additional program forms were introduced with larger studios, and orchestral presentations of music were added, as well as broadcast adaptations of one-act plays.

From 1929 to 1934, there was a tremendous expansion in program types used. New inventions in the field included forms of comedy variety, built around a featured comedian. This is still used today on television, with only slight modifications. Thriller-dramatic programs, including the detective and western adventure, the late-afternoon children's thriller serial drama, the daytime women's serial, as well as the fifteen-minute serial comedy such as "Amos 'n' Andy" became popular. Also, during the same period, public affairs forum discussions, five-times-a-week fifteen-minute news broadcasts, and daytime forms of variety exemplified in the "Breakfast Club" and "Today" shows all set the pattern for most daytime network television programming. In this pattern were country music types of variety programming, and one of the earliest forms of human interest interview programs, named "Day in Court."

In intervening years, a number of new types of programs were developed with the advent of television. We had use of puppet or marionette shows and filmed-for-television cartoon programs for children, plus the introduction of strictly visual sports broadcasts, such as championship wrestling and professional bowling. These news-program format inventions strongly influenced programming at the time.

THE WEARING-OUT OF MUCH-USED PROGRAM FORMS

No matter how successful any single program may be, or how attractive it might be to audiences, it tends to lose freshness and novelty over a period of seasons. It declines in audience attractiveness.

The case of "Dragnet" illustrates this point. In its first season on the air, "Dragnet" had a January Nielsen rating of 35.5. A year later, in January 1953 the rating was 47.2, while in January 1954 it was 54.1. But that was its peak. In January 1955 the rating dropped to a strong 47.2 and in January 1956, to 41.0. In January 1957 the rating was 32.6, and a year later, it registered 26.6. In April 1959 it closed the network run with only 14.6. During the following season, the program was no longer offered on network schedules.

Not only do individual programs wear out, but types of programs reach a high point in popularity, then lose favor over a period of time. Nevertheless, a few remain on the air over several seasons. Senior citizens can remember heavy use of telephone quiz programs by a number of radio stations. This occurred between 1946 through 1950. On this kind of program, the master of ceremonies called a listener by telephone and asked a question and, in the event of a correct answer, awarded a prize.

This form of program disappeared from broadcasting completely, as have man-on-the-street interviews and children's network television programs built primarily around puppets. Even daytime serial programs for women, so popular during the winter of 1940–41, disappeared. In that season, NBC and CBS scheduled a total of fifty-six daytime serial programs each day. All have since disappeared from radio and are now found on daytime schedules for television and cable networks. There has been a resurgence of this type of program on network television, with more sex and human interest appeals in evening serial drama.

Individual programs and types simply wear out, after enjoying high measures of popularity. They lose attractiveness to audiences and are replaced by other program forms.

CHANGES IN AVAILABILITY OF PROGRAM MATERIALS

Material availability has an effect on types of programs provided by radio and television. The decline of vaudeville and burlesque resulted in shortages of new, tried, and proven comedians, particularly before 1940. Although comedy variety programs were widely used on network radio from 1932 until the end of World War II, those programs were built around featured comedians, presented once a week on network television in the past three decades.

However, broadcasters were not able to find sufficiently capable comedians in order to build up those programs. Vaudeville variety programs, of the Ed Sullivan type, were heavily used on network television schedules. But after a year or two, the supply of vaudeville acts ran out. All acts worth using had been aired a dozen times or more on different programs each year.

Now, there are very few programs depending primarily on vaudeville acts, or their equivalent, on television or cable network schedules. "The Ed Sullivan Show" remained almost alone as a representative of this type. In the period between 1952 through 1956, a tremendous number of theatrical feature films, not previously available to television, were released by motion picture production companies for television use. Within a period of three years, every television station scheduled three or four theatrical feature films each day. Since 1956, the supply of new films became less, and today, only a few theatrical feature films of top quality are available. Most were shown more than twelve to fifteen times in every major television market. Home video innovations had an impact on this situation as well.

The result has been a slow decline in total amounts of time devoted to programs presenting theatrical features. Theatrical features rose rapidly to a position of high importance and then declined to a much less important position, simply on the basis of availability of material.

Finally, as recently as 1955–56, television networks had tremendous amounts of open time on daytime schedules. National advertisers were not convinced of the effectiveness of daytime network advertising on television at that time. Since networks provided little in the way of daytime service, each station was forced to provide considerable amounts of local daytime programming that included homemakers' programs, local audience participation shows, daytime sports, and local children's programs. In 1959–60, networks took up far greater amounts of station time, resulting in large proportions of local daytime program offerings, including daytime homemakers' and participation programs, being dropped from station schedules to make way for more innovative network offerings.

CHANGES IN THE INDUSTRY COMPETITIVE SITUATION

In the early 1930s, the U.S. suffered from an extremely serious economic depression. This did not, however, particularly affect the volumes of national advertising on radio networks. National advertisers discovered the effectiveness of network radio as an advertising device. The depression, therefore, did have a profound effect on local advertising.

As of 1932, government reports indicate that of the operating six hundred and fifty to seven hundred commercial radio stations, only forty were able to break even financially. With stations desperate for advertising revenue, several changes took place in local broadcasting situations. Stations facing bankruptcy accepted types of programs that, in better times, they were unwilling to accept.

It was during this period that the "spot announcement" was introduced, as opposed to sponsorship of complete programs by a single sponsor. Secondly, stations allowed direct selling or "hard-sell" commercial announcements. Previously, sponsors were limited to institutional credits, such as "This program has been brought to you, by courtesy of the 'X' company," at the beginning and end of the program.

Thirdly, stations accepted advertising previously rejected. In particular, laxative advertising was introduced. In 1932, '33, and '34, half of all commercial stations carried advertising for "Crazy Water Crystals," which were Epsom salts. Sponsors advertising "Crazy Water" found that the type of program producing most orders was country-western music.

Sponsors of "Crazy Water Crystals" provided country-western entertainers for local programs, for one or two hours each day, on two hundred to two hundred and fifty different stations. Consequently, there was more local programming in the country-western vein than was ever broadcast before, or since, that time.

Fourthly, many stations for the first time were willing to sell time for religious broadcasts, with sponsoring organizations inviting listeners to send contributions to keep the programs on the air. In many cases, those speakers made violent attacks on other religious faiths and preferences. Finally, from two hundred to four hundred stations instituted astrology programs, featuring self-styled "astrologers," who gave advice on everything from business affairs, marriage, and sexual preferences and performance, to styles of dress. This was all done on the basis of a fee, usually $1.00, sent by the person seeking such advice. This particular fee was divided equally between the astrologer and station. With the passing of the worst phase of the depression, some of these programs passed as well, and finally all went off the air.

The second illustration is quite opposite the first. During World War II, many industries engaged in the manufacture of war materials for tremendous profits. To limit the amount retained, Congress imposed the Excess Profits Tax on 90 percent of profits earned by a company in excess of prewar levels. This was a tremendous stimulus for radio network advertising. A company engaged in war production, expecting to produce consumer goods again after the war and wishing to keep its name before the public, had a strong incentive to use the radio networks. For every dollar spent on network program time, 90¢ represented money which might otherwise be paid to the government, in an excess profits tax.

For advertisers at that time, large audiences were less important than "prestige." The result was that during the war, there were more symphony orchestra programs sponsored on national radio networks than at any other time in the history of broadcasting.

Also, during the war, as many as four to eight half-hour documentary dramas dealing with war themes or extolling activities of various branches of the armed services were carried on a sponsored basis by radio networks each week.

CHANGES IN AUDIENCE PREFERENCES FOR DIFFERENT TYPES OF PROGRAMS

This is a highly important factor in determining kinds of programs provided. Audience tastes do not remain constant from one year to the next. Changes may be slight or very substantial within a period of one to two years. A sponsor cannot afford to pay costs of programs, advertising time, local sponsor, or spot announcements to be presented within programs if audiences refuse to listen or watch in appreciable numbers. As a result, audiences decide what types of programs are presented by turning sets to programs they like and enjoy, and not tuning to programs they fail to appreciate. No sponsor or broadcaster can afford to provide programs on the basis of pesonal likes and dislikes if they do not directly correspond to those of the largest number of listeners or viewers.

With respect to programs, there are two major reasons for changes in taste and preference. One has been discussed in connection with the wearing out of programs and forms. There is a decided tendency for audiences, collectively, to be loyal to programs of their choice. With few exceptions, however, this loyalty will not continue for more than a few years.

When a program loses freshness and novelty and impresses an audience with the "same old stuff," audiences lose interest and refuse to watch or listen. This is true with respect to programs of any given type. Today, ten men may like the adult evening "soap opera" type; tomorrow, it may be only four out of ten, or two out of ten. If that is the case, the total number of adult evening "soap opera" types will decrease, with ratings for those programs decreasing as well.

A second reason for changing audience tastes is the change that is taking place in general social or economic conditions. From 1938 to 1941, the threat of American involvement in the European war was high. These same conditions existed prior to the Korean and Vietnam conflicts. But to make the example more graphic, in all these conflicts, there was much concern over the changing fluid international situation. News programs, where experts analyzed the international situation or probable action of our government, attracted large audiences. Therefore, the number of network news programs increased with demand from those audiences.

During U.S. involvement in World War II, Korea, and Vietnam, news concerning progress was anything but good. There were many battle casualties. Our armies made slow progress at the outset. News was the same, from day to day. Even in later stages of those wars, when allied armies were advancing, there was little day-to-day difference in broadcast news. There was still saddening casualty lists, even longer than before.

In the meantime, however, the matter of civilian lifestyles changed. In World War II, there were shortages of food, adequate housing, and ordinary services available in peacetime. The inevitable result was a weariness of war and combat. People didn't want to hear about the war and wanted a means of escape. Listening to network news programs went down steadily. On the other hand, there was a decided increase in escape programs and comedy and thriller drama programs not related to war themes, which provided a temporary escape mechanism from the worries of actual life. These same conditions presented themselves again with the unpopular Korean and Vietnam conflicts, with more graphic broadcast causal relationships developing. Vietnam has been called America's first televised war.

In postwar periods, there are scarcities of many kinds of goods, while prices and taxes remain high; family incomes do not increase in proportion. A result is an unusual interest in big giveaway programs.

These programs show some lucky member of a studio audience winning large quantities of scarce and expensive merchandise. The ordinary American family cannot afford to buy such products, or receive large amounts of money for their purchase. Therefore, listener and viewer interests in such programs are stronger than normal. The number of big giveaway programs on networks increased, while local stations provided their own versions on a smaller scale, of these big give-away ideas.

The local telephone quiz program was highly important during the six to seven years following the end of World War II. It had a great deal of audience interest. We still live in an unsettled world. We have a constant threat of war. We pay high taxes, while the cost of living is continually increasing. Consequently, we're interested in escape, while telelvision programs of the evening serial variety, provide such escapes. Therefore, changes in social and economic conditions produce corresponding changes in programs.

Chapter 3

Radio and Television History—Development of Program Types

Types of programs available to audiences at any given period are strongly influenced by a variety of factors. To understand how programs developed, it is important to have a general knowledge of the manner broadcasting and cable developed, and conditions existing at various stages of that development. The high points of American radio and television should be discussed.

TYPES OF BROADCAST PROGRAMS

Within each period of development, attention should be called to new program forms introduced and to the forms most important within those periods. It is desirable to have a general classification of major types of broadcast programs. Program types are grouped under seven general headings:

a) *Musical Programs*
Concert music—operas, symphony orchestras, concert music live
Musical variety—popular music, large production numbers
Light music—small group or individual pop music
Platter music—music recorded

b)*Variety Programs*
Comedy variety—elaborate production around a featured comedian
Vaudeville variety—series of old vaudeville acts
Semivariety—live music program with talk, comedy, poetry
Amateur talent contest variety—type of "Original Amateur Hour"

Country-western variety—country music and entertainers
Low-budget variety—featured master of ceremonies, interviews, Johnny Carson

c) *Dramatic Program*
Prestige drama—sixty minutes or longer, anthology form, big stars, above average scripting, expensive production
Anthology drama—thirty minutes, different cast and characters in each broadcast
Light drama—same characters, but complete episode each week; home situations, love story interest, not expensive
Comedy drama—situation comedy, built around same character in each broadcast, with supporting regulars, played for comedy, no plot
Informative drama—historical settings, fictionalized with regular plot
Adventure drama-action drama—thirty or sixty minute forms, shot in strange lands, unusual settings, "spy," "foreign intrigue" types included
Detective and crime drama—built around solution of crimes by detective, sometimes around activities of actual criminals
Suspense or psychological drama—built around chillers, lights out, "The Twilight Zone"
Western drama—cowboys and Indians, primarily for children
Adult Western drama—old West scenery, strong in action, unusual settings, emphasis to human interest values, providing adult appeal
Adult serials (two types): 1] fifteen-minute serial form, such as Jack Armstrong, Buck Rogers
2] nighttime serials, "Dallas," "Falcon Crest," et cetera
Women's daytime serials—"General Hospital," "All My Children," "As the World Turns"
Children's drama—dramatized stories for children, including television puppet drama programs

d) *Human Interest Programs*
Interview programs—from "Court of Human Relations," "Man-on-the-Street" interviews with interesting people
Sympathy-arousing programs—"Queen for a Day" with prizes
Studio quiz programs—"Let's Make a Deal," "The Price Is Right"
Telephone quiz programs—call at home, reach by telephone

Comedy-stunt, audience-participation—"Real People"
Teenage dance programs—Dick Clark's "American Bandstand"

e) *Talk-Entertainment Programs*
Panel quiz programs—"What's My Line," "Can You Top This?"
Platter-dialogue programs—sometimes monologue, usually two or three participants, platter for comedy
Children's storytelling—"Tell-Me-A-Story Lady"
Radio broadcasts of sports events—audio play-by-play, television

f) *Talk-Information Programs*
News and commentary—national and local news on important events
Sports news—news about sports, stories about sports figures
Religious talk programs—sermons, religious talks, optional music
Informative talks—general classification, homemakers' programs, farm information, market reports, travel, educational broadcasts in talk forums
Public affairs forums—two sided, or multisided discussions, including reporters' notebooks and roundups, and "Meet the Press" types
Documentary broadcasts—extensive use of dramatic dialogue, essentially talk, inserted dramatized spots, personal ideas
Special events—important happenings described by announcers, radio only

g) *Visual Programs*
Sports broadcasts on television—actual play-by-play, picked up by remote cameras
Actuality broadcasts—special events, such as broadcasts of national political conventions
Actuality broadcasts—travel or documentary illustrating travel forms, old NBC "Wide Wide World" program, documentary would make use of primarily of films

This does not include some programs used on television today. Some were originally prepared for motion picture showing, therefore theatrical feature films are not included. In addition, many forms are difficult to categorize or classify. But the classification includes most forms extensively found on radio, television, and cable today that significantly influence program schedules.

PERIODS IN DEVELOPMENT OF RADIO AND TELEVISION

In the early days of radio, not all forms listed were put on the air by American or foreign radio stations. Some weren't used until after the introduction of television, while others weren t invented until later periods. However, it is convenient to divide programming history into eight visible periods:

1) 1920–26—Early day radio, before radio became commercial
2) 1926–29—Development of commercialization, early operation of networks
3) 1929–35—Period of rapid development, invention of many new program forms
4) 1935–41—Period of prewar stabilization
5) 1941–45—World War II period
6) 1945–52—Postwar period with expansion of radio stations and decline of radio networks
7) 1952–65—Rapid development of television
8) 1965–present—Period of regulation, deregulation, Prime Time Access Rule, 50/50 Rule, Cablevision Rules, cablevision networks

Each period is considered briefly, beginning with the meaning of broadcasting and radio, prior to 1920.

The terms *broadcasting* and *broadcast media* mean dissemination of signals intended to be picked up by a general public, including radio and television signals. This is in contrast with use of radio as a means of point-to-point communications, or citizens' band individual-to-individual communication. Point-to-point radio includes use of radio for communication by ships at sea, aircraft control towers, police departments, taxicab companies, military services, amateur or "ham" operation, and finally citizens' band radio.

Radio communication by the dot-dash method was here as early as 1900. As early as 1910 and 1912, Congress passed laws governing use of radio on board ship. This required ships carrying passengers to be equipped with radio transmitting and receiving apparatus. But this was not broadcasting, as considered in this book.

Broadcasting by radio existed as early as 1910, when Lee DeForrest, inventor of the audio tube in 1907, made modulation of sound and broadcast of the human voice possible. He arranged for a special broadcast of the Metropolitan Opera stars Enrico Caruso and Emmy Dustin.

The broadcast was heard by a primary audience of part-time operators on ships at sea. A good deal of experimental broadcasting on a part-time basis was carried on until American entry into World War I. In the spring of 1917, however, for national security reasons, all experimentation came to an end. It was resumed after government restrictions were lifted in 1919. It wasn't until 1920 that it resumed on a daily, regularly scheduled basis.

First Period: 1920–26

This period marks the beginning of regular broadcasting after the lifting of government restrictions in 1919. Formal broadcasting began on November 2, 1920 when radio station KDKA,[1] Pittsburgh, broadcast the Harding-Cox election returns and inaugurated regularly scheduled daily operations. Major characteristics included amateur, unpaid entertainers, noncommercial operation, low-powered stations, primitive equipment, and listening for purposes of bringing in distant stations, rather than listening to specific programs.

Radio Homes

In November 1920 not more than one thousand American homes had receiving sets. By January 1923, one million radio-equipped homes were represented. By September 1923 two million homes were equipped, representing 8 percent of all homes in the nation. By September 1926 the number increased to five and a half million or 20 percent of all American homes having access to receiving sets.

Stations

In November 1920 only twenty stations, operating on an experimental basis, were on the air. In January 1923 500 stations were on the air, licensed, and operating on a regular basis. In September 1926 only 530 stations were on the air. Throughout the period, no station operated on a full-time basis.

For the first two years, broadcasting time averaged no more than two hours a day, for each station. By September 1926 large stations were on the air for four hours each evening, plus two or three hours during daytime hours. Small stations were on the air for two to four hours daily.

Until October 1, 1922, all broadcasting stations were assigned the same frequency. At that time, the Department of Commerce licensed broadcast stations and assigned frequencies approximating present standard broadcast bands.

Stations were divided into two major classes. Large stations were those owned by electronics companies manufacturing receiving sets including RCA, Westinghouse, and General Electric, together with large newspapers, insurance companies, and retail stores operating on assigned frequencies.

Secondly, small stations were operating with ownership vested in the hands of local businessess, churches, schools, colleges, and private individuals in possession of basic transmitter apparatus, who were on the air for fun. It is interesting to note that through 1926, 135 stations had been licensed for educational institutions.

By autumn 1926 forty large stations operating in large metropolitan areas, together with small stations using very low power, were operating on a regularly scheduled basis.

By the end of 1922, less than twelve stations used more than fifty watts, not kilowatts, of power. Most increased power to the fifty-watt peak. Small stations divided time on the same frequencies, in the same cities. Most used ten to fifty watts of power. By 1926, eight of the large stations were using 5,000 watts of power. All experimentation leading to developments in radio were carried on by large stations, especially those owned by electronic concerns.

Networks

No permanent networks existed in September, 1926. Large stations occasionally hooked up in a temporary network, and even a coast-to-coast broadcast of a political rally occurred in autumn, 1924. The period 1924 through 1926 saw several large stations linked to an informal network, extending from Washington to Chicago, for occasional broadcasts.

Equipment

Early microphones were very insensitive. Studios were too small to accommodate orchestras. Not until the end of the period did large stations begin to use more sensitive ribbon microphones or build larger studios.

Commercial Operation

Virtually all commercial operations were nonexistent. The first commercial program was broadcast by WEAF, now WNBC, in New York, October, 1922. In January 1926 this same station had more commercially sponsored programs a week, but the revenue did not meet the operating costs! Other large stations carried occasional commercial programs. For small stations, in September 1926 sale of commercial time was simply unknown and financially impossible.

Programs

With stations having no revenue, programs were those that could be provided without cost. Talent was unpaid, and announcers in small stations worked for fun, while at large stations, announcers were employed on a part-time basis.

Through 1924, the only forms of programming used included light music from studios. Remote pickups of small orchestras, dinner music, or chamber music, and various storytelling for children, as well as talk programs, centering upon informative and religious talks, all originating at studios, were introduced.

By September 1926, large city stations presented musical variety programs from enlarged studios, with concert music. There were no dramatic, variety, or audience participation programs carried on any station. A few experimented with occasional remote broadcasts of sports events; however, no sports programs were broadcast on a regular basis.

Types Enjoying Greatest Popularity

light music
informative talks
religious talk programs
story telling for children (including Christmas)
early type musical variety (limited)
concert music (limited)

Second Period: 1926–29

The period was highlighted by the serious beginning of commercial broadcasting, organization of national radio networks, increased power and operating hours for stations, as well as shifts of control of broadcasting from the Secretary of Commerce to the newly organized Federal Radio Commission. Prior to 1927, all broadcasting stations were licensed by the Department of Commerce, under the Radio Act of 1910. The Commerce Department was empowered to license those stations not satisfied with frequencies assigned to them.

Stations, however, challenged the licensing power of the Department of Commerce for broadcast stations in federal courts. The courts ruled that the Department of Commerce had no authority over broadcasting stations, as opposed to those stations operating on a point-to-point communication basis. Congress subsequently passed the Radio Act of 1927, creating the Federal Radio Commission, and gave the commission power to license both broadcasting and point-to-point communication stations prescribing operating power, assigning frequencies, and determining hours each station could operate.

Radio Homes

By the autumn 1929, there were approximately nine million homes with radio sets, compared with five and a ½ million three years earlier. This was approximately 30 percent of all homes in the United States.

Stations

The number of stations increased by less than one hundred. In September 1929 there were six hundred stations, compared with 530 stations in 1926. In September 1929 a third of this number were new stations. New regulations provided for minimum operating hours and careful adherence to exact authorized frequency. The net result drove more than one hundred and fifty operating in 1926 off the air. This included one hundred educational stations that lacked financial ability to meet new FRC guidelines and requirements.

A trend toward increased station power continued. In 1927, WGY, Schenectady, New York, increased power to 50,000 watts. This was the first station to use such high power. By 1929, five other stations were using similar power. Large stations used at least 5,000 watts of power, while small stations also moved into upper categories, with no more than forty stations using less than 100 watts of power.

By fall 1929, less than one third of all stations operated on a full-time, eighteen-hour-a-day basis. No station provided programs for broadcast five days a week, at the same hour. Daytime programming, like that at night, consisted chiefly of once-a-week programs.

Networks

By the end of 1927, three networks, later to become nationwide, came into being. NBC commenced operations with its Red Network in December 1926. The same network company commenced operation of a second system, the Blue Network, later to become ABC, in January 1927. CBS commenced operations in September 1927, with service limited to stations in cities east of the Ohio and north of the Missouri Rivers. The NBC-Red Network inaugurated coast-to-coast service in December, 1928.

Station Equipment

Most stations increased the size of their studios, while others provided studios large enough to accommodate audiences. Stations carefully separated audiences from performing areas by sheets of plate glass. Condenser and ribbon microphones were in general use, with ribbons favored because of directional capabilities. The World Broadcasting Company introduced transcription libraries. Subscribers to World Service were provided with thirty-three-and-a-third RPM turntables. Transcribed or recorded music was used only for backgrounds or transitions.

Commercial Operation

Of the nearly six hundreds stations operating in 1929, forty operated on a noncommercial basis. The noncommercial stations were primarily educational or were operated by religious organizations. All others sold time, although in 1929, half did not have enough revenue from sale of time to pay operating expenses.

A few engaged in direct selling of merchandise, with listeners sending orders and money directly to the station. Most income for stations and networks alike came from sale of time to advertisers of sponsored programs. During the 1928–29 season, total revenue from sale of time amounted to about $14 million with 80 percent of the total going to networks. Advertising was entirely institutional, with sponsored programs identified as "presented by the courtesy of" the sponsor, with no direct-selling commercials permitted. Spot announcements were not introduced to the industry at this time, as has been popularly suggested by many writers.

Programs

All sponsored programs were either thirty minutes or sixty minutes in length and were presented on a once-a-week basis. There were a few five-times-a-week programs in daytime schedules, but the majority of daytime programming was scheduled on a once-a-week basis.

A list of program types used on networks during the periods considered shows the network influences. Since networks came into existence during this period, all networks were concerned in programming decisions.

New Network Program Types

General variety—('26–27) "Eveready Hour" (different type of material each week)
Country music-minstrel—('26–27) "Dutch Masters Minstrels"
Concert music—('26–27) "Atwater Kent Hour," "Cities Service Concert"
Musical variety—('26–27) "A & P Gypsies," "Cliquot Club Eskimos"
Light music—('26–27) "Jones & Hare," "Trade and Mark" (both song and platter)
Light, love-interest, home drama—('26–27) "True Story Hour"
Informative drama—('26–27) Biblical dramas, "Great Moments in History"
News commentary—('26–27) Frederick Wm. Wile, H. V. Kaltenborn
Public affairs forum—('27–28) "Voters' Service"
Religious talks—('26–27) Reverend S. Parkes, Rev. Dan Poling
Informative talks—('26–27) Betty Crocker, Dr. Royal S. Copeland
Entertainment talks—('26–27) "Cheerio"

During the 1929–30 season, networks provided fifty-six hours of evening and thirty hours of daytime programs each week, including combinations of Sunday afternoon and evening variations on similar themes.

For evening totals, forty hours were devoted to music, five hours to variety and drama, and six hours to talk programming. Of daytime totals, nine hours were devoted to music and twenty-one hours to talk. In addition, there was one fifteen-minute dramatic program broadcast during daytime hours.

Local programs used the same proportions as those on networks. Important forms included musical programs, especially light music, and talk.

For networks, the period saw use of only a limited number of variety programs, with an early form of "The Eveready Hour," presenting a different general type of material each week. One week, there was a dramatization adapted from a short story, while the next week's presentation included a debate with parliamentary rules of order.

"The Dutch Masters Minstrel" was the sole representative of the country/western music or minstrel classification. Concert music filled a large proportion of network schedules, with one third of all time devoted to music, including symphony orchestras, operas, and chamber music. Light music took the form of presentations of two-man song and platter teams. There were no network broadcasts of straight news. News commentary was presented on a once-a-week basis, with single commentary programs scheduled each week on each particular network.

Third Period: 1929–35

During this period, network advertising became firmly established, with network revenue increasing steadily. Individual stations were adversely affected by depression conditions.

Local advertising had not become important yet and was severely curtailed by depression conditions. In 1933, nine out of ten stations operated at a loss. As a result, advertising standards were lowered on both stations and networks. Types of broadcast programs accepted were considerably below standards that broadcasters previously had accepted. This was the period, however, of the greatest invention of new programs in the history of broadcasting. During this period, the Federal Communications Act, 1934, as amended, was enacted into law by the Congress. It replaced the Federal Radio Commission as the regulatory body for broadcasting.

Radio Homes

In spite of the depression, the number of radio-equipped homes continued to increase. By September 1935, there were more than twenty-two million radio homes in the nation. There was a drastic cut in the average price of radio receiving sets, from $120 in 1929 to $40 in 1935. Also, two and a half million families had automobile receiving sets in September 1935.

Radio Stations

The number of radio stations remained almost unchanged. Statistically, in September 1935 approximately six hundred stations were on the air. Power and operating hours increased. In 1935, nine out of ten stations were licensed for full-time operation and were on the air from sixteen to eighteen hours daily. Of the six hundred stations, thirty used power of 50,000 watts, and nearly one hundred and fifty were broadcasting with 5,000 watts of power.

Networks

Networks literally became big business entities in this period. By 1935, each of the three original networks operated on a coast-to-coast basis, with affiliates ranging from eighty to one hundred and twenty stations each. In addition, a fourth network, the Mutual Broadcasting System, was organized in 1934. Starting with only four stations in 1934, it had sixty affiliates by September 1935. Programs and revenue, however, restricted Mutual from becoming a serious rival to the three longer-established network operations.

Station Equipment

No major change took place in the type of equipment used. During this period, glass windows separating performers from studio audiences were removed.

Advertising

In spite of the depression, total advertising revenue from radio rose steadily. For 1935, industry revenue totaled $80 million or five times the estimated amount for 1928–29. Half of this total represented payments by advertisers to networks. About $14 million was paid to stations for national nonnetwork spot announcements or advertising, and $26 million came from local advertisers. However, during 1933, 90 percent of all stations operated at a loss.

Depression conditions had a decided effect on types of advertising carried. First, networks and stations accepted selling commercials, while through 1926–29, virtually all radio advertising was institutonal. Second, advertising was accepted both by networks and stations for products and services not previously considered acceptable, including laxatives, deodorants, funeral homes, astrologers, and religious organizations soliciting contributions. Third, a large number of stations engaged in direct selling, or cost-per-inquiry advertising, with listeners sending purchase orders directly to stations.

Stations generally sold spot-announcement time to advertisers. Some announcements were inserted in station breaks between programs; others were carried in participating programs; and in most cases, homemakers and farm information programs carried all these announcements.

Programs

As noted, this was a period of rapid change in types of programs provided, with more new forms introduced on national networks than in any other period of American broadcast history.

New Network Program Types

Comedy variety (early form)—('29–30) "Cuckoo Hour," "Nitwit Hour"
Comedy variety (regular form)—('30–31) "Eddie Cantor Program"
Vaudeville variety—('30–31) "Rudy Vallee Varieties"
Semivariety—('29–30) "General Electric Hour," "Symphony with Floyd Gibbons"
Amateur contest variety—('34–35) "Matopma"; "Amateur Night"
Country western variety—('33–34) "Corn Cob Pipe Club," ('34–35) "National Barn Dance"
Low-budget variety (daytime)—('33–34) "Breakfast Club"
Prestige drama—('29–30) "Radio Guild Drama" (daytime, sustaining)
Anthology drama—('30–31) "First Nighter"
Fifteen-minute drama—('30–31) "Amos 'n' Andy"

Adventure drama—('29–30) "Empire Builders"
Detective-crime drama—('29–30) "True Detective"; ('30–31) "Sherlock Holmes"
Western drama—('33–34) "Box XX Days," ('34–35) "The Lone Ranger"
Adventure serial drama for children—('31–32) "Little Orphan Annie"
Women's daytime serials—('31–32) "Clara, Lu & Em;" "Myrt & Marge" (evening)
Children's drama (fairy stories)—('33–34) "Adventures of Helen and Mary"
Human interest interviews—('33–34) "Alexander's Court of Human Relations"
Platter-dialogue—('30–31) "East and Dumke's Sisters of the Skillet"
Sports news—('34–35) "Eddie Dooley's Sports Talks"
Broadway and Hollywood gossip—('32–33) "Walter Winchell"
Five-times-a-week new broadcasts—('30–31) "Lowell Thomas"
News documentary drama—('31–32) "March of Time"

With invention of so many new network forms, there were material changes in the proportion of programs of different general types carried on networks. During the winter seasons of 1934–35, of approximately one hundred and thirteen hours of evening network programming each week, the major portion, fifty-two hours, was concentrated with musical programs of the musical variety type, while one-third consisted of the concert music distinction.

More than twenty-one hours each week were devoted to dramatic and variety programs. Evening and Sunday talk programs filled seventeen hours per week of network time, and a relative "newcomer," human interest programming, accounted for approximately two hours of evening program time per week.

During the daytime, networks provided eighty hours of programming each week, compared with thirty hours during the 1928–29 periods. Musical and talk programs, the latter including the entertainment type, together with informative talks accounted for about twenty-two hours each. Variety, with the "Breakfast Club" in particular, accounted for ten hours, with human interest accounting for five fifteen-minute periods. The most important increase was in the field of drama, filling twenty-five network daytime hours each week. Half of the total consisted of women's daytime serial dramas, and another four hours consisted of children's adventure serials. Although the five-times-a-week Lowell

Thomas program went on NBC's Blue Network in the autumn of 1930, news was still not an important factor in network programming at any time during this period.

Local Programming

The period began with major innovations in local programming, but not of the same types which took place on networks. During most of the period, stations were in serious financial difficulties. To increase station revenue, program types were added, and their use expanded to specific objectives of bringing in additional revenue.

For example, during this period, at least one-third of all stations sold time. This was in addition to any free time granted for religious programs, allowing those religious broadcasters to solicit contributions from listeners. An equal number of scheduled programs featured self-styled astrologers who invited listeners to send in dollars for advice. This came to a halt when FCC action was taken against the astrologers.

As many as one-half of all stations featured programs of local country western music, and in most cases, sponsored it with a laxative advertiser, Crazy Water Crystals. A small number featured local live orchestras appearing free in exchange for on-the-air promotion of future dance dates. A majority of stations introduced various types of programs adapted to the need of participating sponsorships. There were homemakers or farming programs, while others provided programs of recorded music. Many stations introduced regular news broadcasts, usually five minutes in length, with news lifted directly from columns of local newspapers.

Fourth Period: 1935–41

This was the pre–World War II period, which sparked recovery from the depression. The period involved expanding revenue for radio networks and stations as well as widespread concern over dangers of involvement in a "European War," during the latter half of the period.

Radio Homes

Total numbers of radio homes increased by September, 1941, to over twenty-nine and a half million homes, or roughly 87 percent of all homes in the United States. In addition, by 1941, some eight million auto sets were in use.

Radio Stations

By September 1941, the total number of stations increased to more than eight hundred and fifty. Forty-five operated with 50,000 watts of power, and an additional one hundred and fifty with 5,000 watts of power. New stations were authorized in the low-power classification. For low-powered stations, 250 watts had become a standard minimum. All stations operated on a full-time basis and provided program service for eighteen hours each day.

Radio Networks

The same four major network companies, NBC-Red, NBC-Blue, CBS, and Mutual, provided network service to seven hundred of approximately eight hundred and fifty affiliated radio stations. The first three companies had one hundred and thirty to one hundred and fifty affiliates each. Mutual Broadcasting had three hundred affiliates by 1941, but all were low-powered stations. By the latter part of the period, CBS became a strong rival to NBC-Red for dominance in the network field. NBC-Blue was in third position in annual revenue, and Mutual ran a poor fourth.

Advertising

Total radio revenue from sale of time to advertisers doubled over the period from $80 million in 1935 to $180 million for the year 1941. Of this total, $82 million went to networks; another $46 million went to stations from national spot advertising; and $52 million represented local advertising. There were substantial increases in volume of national spot nonnetwork advertising, by national advertisers. While network revenue doubled between 1935 and 1941, the amount spent for national spot advertising was three times as great in 1941 as in 1935.

With substantial increases in volume of local advertising, network advertising for the first time accounted for less than half the total revenue of networks and stations. Accompanying this was a substantial increase in use of spot announcements, as opposed to purchase of time on local stations for complete programs. This applied to local stations, with no participating programs carried on networks. With substantially increased revenue, all stations were operating in the black, and large stations as well as the 5,000-watt variety were earning relatively high profits.

Television

While television was by no means important during this period, it should be covered, nevertheless. Prior to 1941, twelve to fifteen experimental television stations had been authorized by the Federal Communications Commission. In 1941, the FCC authorized regular commercial television broadcasting, and on July 1, 1941, NBC's television outlet in New York City, WNBC,[2] started commercial operation, with different sponsors for its programs.

Before the end of 1941, three additional stations were licensed for commercial operation. Our entrance into World War II brought commercial operations to an end. As of 1941, the total number of television sets in existence in the United States did not exceed 10,000.

Programs

The major characteristics were an expansion in amount of time devoted by stations to network programs during daytime hours. A tremendous increase in use of news programs on networks and local levels, rise of name network commentators, and major expansion of local participating programs in the area of recorded music all highlighted the era. Few new programs were introduced, however, and accordingly, the period was characterized by continued development of existing program forms. There was an introduction of quiz and human interest types, with rapid expansion in the use of these program types.

For programs generally, starting in 1935, services of national newsgathering organizations, such as Associated Press, United Press, and International News Service, were made available to networks and stations after resolution of legal entanglements referred to as "The Press-Radio War." This arose from unauthorized use of press-service news by radio stations.

Broadcasters were involved in still another legal squabble at the same time. Beginning with the 1938–39 period, the American Society of Composers, Authors, and Publishers held copyright authority to available music; consequently, ASCAP demanded networks and stations pay fees for rights to use music ranging up to 5 percent of their gross revenue from sale of time to advertisers. For a period of several months, no ASCAP music was performed on radio. During this period broadcasters themselves set up a competing licensing organization, entitled Broadcast Music, Incorporated. An industry committee, however, worked out more favorable arrangements with ASCAP, and since 1939, both ASCAP and BMI licenses have been held jointly by most stations.

New Network Program Types

Evening network program offerings totaled 120 hours per week with only slight increases over the number provided in 1934–35. But program distribution was considerably different in 1940–41. Musical programs accounted for thirty-two weekly hours, down from fifty-two hours a week in 1934–35. Variety programs accounted for seventeen hours, down from twenty-one hours. Use of drama increased slightly to twenty-four hours each week in 1940–41, and various types of human interest programs filled fifteen hours of evening network time each week, as compared with only two hours in 1934–35. Talk programs increased to twenty-seven hours each week, compared with seventeen hours in 1934–35, with most of the increase from a weekly total of twelve hours a week of news broadcasts and forum discussions of public issues.

Daytime network programming showed even greater changes. By 1940–41, daytime network hours had increased to 147 hours per week with an increase of 80 percent over 1934–35. There was a slight decrease in actual program hours devoted to daytime variety and musical programs. Similarly, some decrease occurred in the number of daytime informative talk programs other than news. No news programs were included in daytime schedules either in 1935–36, or in 1940–41. In both years, the amount of human interest programming offered during daytime hours each week was the same, with a little more than an hour.

In the dramatic program field, there was a tremendous increase in 1940–41. There were no less than sixty different serial dramas on the air every day. Fifty-two serial dramas were scheduled by two of the four networks, NBC-Red, and CBS. Daytime serials accounted for seventy-five hours a week of network time. Children's adventure serials accounted for an additional ten hours, and the two forms combined made up for more than half of all daytime network programming.

Local Programming

Four major changes took place in the type of programming provided by individual stations during the period. First with heavy expansion of daytime programming by networks, more time was devoted, by stations, to network offerings in 1940–41 than had been the case in 1934–35. This left less time for locally originated programs. Second, local stations, like networks, greatly expanded the amount of news programs provided. This was, in part, a result of tremendous listener interest in news and availability of service to stations from major news-gathering agencies, including Associated Press, United Press, and International News Service. The average station by 1940–41 offered from three to five locally originated fifteen-minute news programs a day. This was in addition to the news programs provided by the networks.

Third, there was an expansion in use of programs of recorded music, whereas prior to 1935, and as late as 1938, many broadcasters felt the public would not accept recorded music. That feeling gradually disappeared as a result of successful use of recordings on several big-city stations. By 1940–41, stations affiliated with networks offered an average of two hours a day of recorded music, and in most cases, participating programs were vehicles whereby spot announcements for national and local advertisers could be inserted.

Stations without network affiliations were broadcasting six to ten hours of recorded music each day. The fourth change was a result of expansion of network programming. With news and recorded music, allotments of time were devoted to these programs, while other forms were used less. Live music, aside from the country western variety, almost disappeared from the field of local origination in this period. Even country western programs themselves were down by 1940–41, with no more than one-third the number carried in 1934–35. There was also a reduction in the amount of time devoted to locally produced informative talk programs, such as farm broadcasts and homemakers programs, although most stations retained programs in each of these categories.

The growing success of human interest and interview programs on networks led to the introduction, on several stations, of locally produced man-on-the-street broadcasts. These programs were not important, but stations using the form were broadcasting only one such fifteen-minute program each day. The form, unknown in 1934–35, was widely used by local stations in 1940–41.

Fifth Period: 1941–45

The nation shifted on December 7, 1941, from a threat of war to actual involvement in World War II. The war imposed many hardships on broadcasters, as it did upon those engaged in other occupations. Electronic concerns shifted to war production, resulting in broadcasting equipment, tubes, and electrical engineering apparatus in very short supply.

Costs advanced, while employees were drafted into military service. On the other hand, both network and station revenues were strikingly greater than in the prewar period. In 1944–45, it was 70 percent above levels of revenue in 1940–41. Some of this increase was the result of inflation, but even so, 1941 to 1945 was the big money-making period for American radio.

Radio Homes

Virtually no radio receiving sets were produced from mid-1942 through the end of the war; nevertheless the number of radio homes increased. By 1945, there were 34 million sets, compared with less than 30 million units in 1941. The sets used in additional homes were primarily second sets, which in 1941 had been in homes of parents or relatives. By no means were all sets in the 34 million homes in working condition in 1945. Set repair and tube replacements were extremely difficult to obtain during the war. The number of radio-equipped automobiles in 1945 was only six million compared with eight million in 1941.

Radio Stations

The number of radio stations increased by the end of 1945 to 940, which represented ninety more than in 1941. The additional stations had been authorized before the beginning of the war and, therefore, with transmitters and equipment finally secured, construction started.

Increases in operating power were few in number, due to difficulties in securing transmitters. At the end of 1945, there were forty-five 50,000-watt stations, and one hundred and seventy-five 5,000-watt stations.

Radio Networks

Only one major change occurred in the status of networks during this period. The "Duopoly Order," from the Federal Communications Commission, became effective in 1941. This order prohibited ownership of more than one network by a single network operating company. NBC was forced in 1942 to sell its Blue Network to a different corporation. The name of the network was changed to the American Broadcasting Company.

Few stations were added to affiliate lists or each of the networks during the period. Throughout, NBC (formerly NBC-Red) remained the leading network in program popularity and income. CBS was second, while ABC represented a weak third, and Mutual was an even weaker fourth.

Television

Television activities were brought to a complete standstill during the war years. There were not more than 10,000 receiving sets, and half of those were not in working condition by 1945. Although the Federal Communications Commission authorized commercial operation in 1941, there was no incentive for sponsors to buy time on television stations. During the war period, existing stations operated only on a token basis, broadcasting for two to four hours each week.

In 1945, only seven television stations had been licensed for commercial operation, including three in New York and one each in Philadelphia, Schenectady, Chicago, and Los Angeles.

Radio Advertising

In spite of the war, advertising revenue increased tremendously to a total of $310 million in 1945, compared with $180 million in 1941. Of this 1945 total, network revenue accounted for $134 million, national spots for $76 million and local advertising for $100 million. One measure having great effect was Congressional enactment of a law imposing a 90 percent tax on excess profits, especially war industries. As a result, a company whose income was subject to tax could buy advertising at a net cost of ten cents on the dollar.

Even industries engaged entirely in war production, with no consumer goods, advertised heavily to retain competitive position when the war ended. Looking to the future, company executives realized they would need the goodwill of consumers. Therefore, several industrial concerns were encouraged to continue advertising efforts by the excess profits tax.

Programs

Programs on networks and those locally originated reflected the war situation. Throughout the entire period, there was strong emphasis on news with human interest programs using men or women in the armed forces as participants. With many sponsors having nothing to sell, there was a greater use of institutional commercial announcements than in either the prewar or postwar periods.

Network Programming

No new network program forms appeared during the war period. However, there were some changes in use of certain program forms. News and commentary programs were extremely popular during the early stages of the war. Particular emphasis was placed on commentary; therefore, fewer than half of all news programs were straight news, with emphasis upon the commentary format.

The total number of news programs—one hour per week devoted to news broadcasts during evening and Sunday hours on networks—was twice as great in 1944–45, compared with eleven hours a week in 1940–41, showing the direction of news programs. By the 1943–44 season, a certain amount of war-weariness was evident in the listening public. While there were more news programs, amount of listening to network news decreased. This same war-weariness, and an evident desire for escape, produced increases in listenership to escape-type programs, including comedy variety, comedy drama, and thriller drama. All programs in these classifications had higher ratings than in the prewar period, and the amount of time given to these forms increased proportionally.

During the winter of 1944–45, networks scheduled eight hours of comedy variety and comedy drama, with fourteen hours of thriller drama also provided each week. The corresponding number of hours for the three types in 1940–41 was four and a half, four, and nine respectively. Another form showing increased use during the war period was information drama. There were no evening programs of this type on networks in 1940–41, but nearly five hours of such programs per week in 1944–45. Total time devoted to all evening variety forms, including comedy, in 1944–45 was approximately nineteen hours per week, virtually the same as in 1940–41.

In musical programs, thirty-four hours per week were offered, compared with thirty-two hours in 1940–41. To all types of dramatic programs in 1944–45, there were thirty-two hours per week provided, as compared to twenty-four hours in 1940–41. There was a decrease in use of audience participation and talk programs other than news. In 1944–45, quiz and human interest programs totaled slightly over ten hours per week, compared with fifteen hours in 1940–41. In the final year of the war, talk programs other than news used only fourteen hours each week, compared with twenty-five hours in 1940–41.

Two facts deserve special mention. War industries buying evening network time were interested in prestige programming, while ratings were of secondary importance. More symphony orchestras were presented on a sponsored basis than at any other time in network history. Similarly, many war industries presented informative dramatic programs and, in some cases, documentaries dealing with contributions of various military services to the war effort.

Secondly, a decrease in quiz and other types of human interest programs occurred. This is difficult to understand, since at the same time, daytime use of such programs increased substantially, while all types of audience-participation programs made use of men from the armed services.

For daytime network programming, time devoted to news programs increased to thirteen hours a week in 1944–45, compared with five hours in 1940–41. There was a similar increase in daytime use of nonthriller and nonserial dramatic forms. In 1944–45, these programs totaled fourteen hours a week, with more than four hours used for informative drama, as compared with only one hour each week in 1940–41. A similar increase was evident in daytime quiz and human interest programs. From only one hour each week in 1940–41 to eleven hours in 1944–45.

Daytime network use of variety programs added twelve hours a week, while daytime musical programs reached eighteen hours a week, remaining vitually unchanged from the 1940–41 level.

Major reductions were registered in nonnews talk programs: thirteen hours per week in 1944–45, compared with twenty-three hours in 1940–41. In time devoted to daytime serial drama, forty-nine hours were offered each week in 1944–45, compared with seventy-five hours in 1940–41. Daytime serials and daytime information programs undoubtedly had passed their peak of popularity and were on the decline.

Local Programming

Aside from greater emphasis on materials related to war, there was little change in local programming during this period. Some stations added commentators to news staffs and attempted to rival networks in providing news commentary programs. Many stations arranged for presentation of recorded interviews with men in military service who came from communities served by the station. A third to half of all stations made use of local human interest programs, including interviews, quiz, and audience participation. These were virtually the only changes in the lineup of locally produced programs during the war period.

Sixth Period: 1945–52

This was a highly important period in the history of American broadcasting. An enormous increase in total number of radio stations, construction on more than one hundred television stations, and the beginning of a shift in importance from radio to television, especially on the network level, began to develop into a pattern and trend.

Radio And Television Stations

Following the close of World War II, the Federal Communications Commission took two actions that had a striking effect on the radio industry. First, the FCC reduced required minimum distance between any two stations on the same AM frequency, while authorizing construction of daytime-only AM stations on frequencies formerly reserved for use by clear-channel, or 50,000-watt, AM stations.

Secondly, the FCC opened a substantial band of frequencies for FM, frequency modulated broadcasting and implied strongly, that there was a distinct possibility within a few years that all radio broadcasting would be shifted to FM. The result was a tremendous increase in numbers both of FM and AM stations. AM stations increased from nine hundred and forty in December, 1945, to nearly twenty-four hundred in autumn of 1952. At the same time, power increases were granted to many AM stations, which was impossible during the freeze on equipment during the war. By September, 1952, there were ninety AM stations operating with 50,000 watts of power, and three hundred and fifty operated with 5,000 watts of power. Some of the low-powered stations, however, were included in the seven hundred stations authorized to operate only during daytime hours.

During the first few years of operation, many new FM stations were programmed independently, although six hundred of the six hundred and fifty on the air in 1952 were licensees of AM stations, operating in the same markets and from the same studios as their AM counterparts. Within a year, all FM stations connected with AM stations were adopting policies of duplication. This was a simultaneous effort to program FM stations with AM counterparts. Their owners had not been successful in selling enough time on FM stations to pay operating costs, and duplication of AM service on FM stations certainly cut those costs.

Television stations came into existence more slowly. At the close of World War II, not more than seven television stations had been licensed for commercial operation. Although many broadcasters believed in the future of television, construction of a television station cost from $750 thousand to $1 million. This was a great deal of money to risk, when commercial television had no outward opportunity to prove itself. Furthermore, at the end of the war, fewer than ten thousand television receiving sets were in existence, and until the Federal Communications Commission made a decision as to channel allocations to be used for television, manufacturers did not dare attempt to produce sets. Up to October, 1948, 108 television stations, all located in larger cities, were authorized by the Federal Communications Commission.

During that month, the FCC declared a freeze on granting additional licenses until it had an opportunity to consider color requirements and address educational television concerns, as well as to provide additional facilities for smaller cities. The freeze was not lifted until April, 1952. In the next few months, a number of additional stations were authorized, but none were on the air until September, 1952. As of September 1952 there were exactly 108 television stations in operation, occupying channels on the VHF band, and all located in major metropolitan areas.

Radio-Television Homes

After electronics companies were able to shift from war to peacetime production, they turned to radio set production to offset a lack of production during the years of war. In 1947 alone, twenty million sets were produced, with retail value nearly $1 billion. Many replaced old sets used during the war. Many sets became second or third sets in the same home, and naturally, large numbers were used to equip new homes, as a result of the housing boom following World War II.

As of autumn 1952, the number of radio homes had increased to forty-six million, from 95 to 97 percent of all American homes, while the number of radio sets in automobiles reached twenty-five million. However, the auto sets and 80 percent of all home sets were equipped to receive only AM signals. The number of homes with sets capable of receiving FM signals was not more than six to eight million.

The number of television homes increased more slowly because few families were willing to spend $500 or more to buy a television set, until one or more television stations were in operation in the home community. At the end of 1948, there were not more than one million television-equipped homes. In 1949, four million were in U.S. homes, and in 1950, there were ten and a half million sets available. In the autumn of 1952, more than twenty million sets were located in U.S. homes. This accounted for 42 percent of all American homes.

Radio Networks

The same four networks continued to provide service to radio stations throughout the period, with an increased number of affiliated stations. By 1952, NBC and CBS were serving two hundred AM stations each. ABC had two hundred and seventy-five affiliates, and Mutual nearly four hundred. One major shift in network status took place shortly after the war. CBS, for years running second in popularity and volume of business to NBC, bought a number of NBC's most popular entertainers, including comedians, and placed them on their fall programs.

Accordingly, CBS replaced NBC as the leading network; NBC was a very close second; ABC ran a poor third; while Mutual was an extremely weak fourth.

Also during this period, a new coast-to-coast radio network came into being. The Liberty Broadcasting System started operations by providing recreated baseball broadcasts to fifty stations. Later it expanded its activities to provide a few hours of program service daily to three hundred stations. Unfortunately, after two years of operation, Liberty was forced into bankruptcy in 1950. In 1952, eleven hundred stations previously receiving Liberty's network service began receiving service from the four established networks. At this time, some thirteen hundred stations, including six hundred daytime only stations, operated as totally independent radio stations.

Television Networks

Television networks came into existence almost before any television stations had been constructed. By 1946, three of the existing radio networks, CBS, NBC, ABC, and one new aspirant, DuMont, were busily engaged lining up affiliates, while inducing radio affiliates to build television stations, so they could be affiliated with televisions as well as radio network operations.

As of 1952, all television stations, aside from fourth stations in nine four-television stations markets, together with the fourth, fifth, sixth and seventh stations in New York and Los Angeles (both cities had seven television stations each) had network affiliation contracts. Many were in markets with fewer than three stations, with two or more of those being affiliated with the networks. In competition for affiliates, DuMont found itself at a serious disadvantage and, as a network, never became firmly established. It attempted limited network programming in 1951–52, but ultimately gave up the venture. The first year of operation for television networks was in the season 1948–49.

Advertising and Economic Conditions

The volume of radio advertising continued to increase throughout the period, in spite of the emergence of television. For the calendar year 1952, radio time sales totaled $470 million, with $100 million going to networks. This was a decrease from the all-time high radio networks reached with $141 million in 1948. One hundred and twenty-three million dollars came from national spots, and $240 million from local advertising. This total was split among more than three times the number of stations, including FM stations, which were in operation at the end of World War II. The majority of new AM stations and all FM stations were virtually operating at a loss. The trend toward reduced radio network earnings was already strongly evident when national advertisers started shifting expenditures from network radio to network television.

Television revenue from sale of time had increased rapidly in less than four years of television operation. In 1949, total television revenue totaled only $27.5 million. In 1950, it reached $91 million; in 1952, it brought $209 million; and for the year 1952, $283 million, with $136 million of the 1952 total going to networks. This was a larger amount than network radio received at the same time. Also, $80 million came to stations from national spot advertising, while $65 million represented local advertising.

Although television advertising was increasing at a sensational rate at that time, television networks and stations operated at a heavy loss during these early years. By 1952, not more than half of all television stations were earning a profit. As might be expected, the types of programs provided by radio were greatly affected by the new economic conditions.

Radio Network Programming

Several changes occurred in programming provided by radio networks. Evening variety and musical programs decreased even more significantly, as did evening quiz and audience participation programs. News and evening talk programs held at the same level. Use of evening dramatic programs increased strongly over 1944–45. In 1951–52, networks offered forty-six hours a week of dramatic programming during evening hours and on Sundays, compared with thirty-one hours per week in 1951–52. The networks had no less than twenty-five hours each week consisting of thriller drama programs.

During the daytime, there was a striking increase in low-cost variety, as well as quiz and audience participation. Since it was difficult to differentiate between the two forms, these two types accounted for fifty-nine hours of daytime programming each week, compared with twenty-three hours in 1944–45. Daytime serials decreased slightly, to forty-four hours pers week. Other daytime dramatic programs decreased to fourteen hours per week, as compared with twenty-four hours in 1944–45.

During the period, daytime adventure serials for children were almost completely replaced by thirty-minute adventure dramas. Each program in a series was built around a single hero, with each thirty-minute episode constituting a complete story. During this period as well, networks for the first time permitted the use of recorded music. As of 1951–52, networks scheduled one and a half hours of platter music programs at night and an additional seven and a half hours each week during daylight hours. The really important change was in sponsorship of network programs. In 1944–45, 90 percent of all radio network programs were sponsored. By 1951–52, not more than 50 percent of all evening network hours and a small proportion of daytime hours had national sponsors. Eight to 10 percent of all network programs were cooperatives, fed to affiliated stations for local sponsors, and 40 percent were broadcast on a sustaining basis.

New Radio Network Program Types

Platter music—('47–48) "Martin Block Show," "Paul Whiteman Record Program"
Telephone quiz—('48–49) "Stop the Music"
Low-key detective programs—('49–50) "Dragnet"

Local Radio Programs

Local programming was strongly influenced by three factors. First, the tremendous increase in the number of new stations left more than twelve hundred stations, as of 1952, without network service. Secondly, the drastic decrease in quantity of sponsored network programming left affiliated stations with more hours of program time to fill locally than had previously been the case. Third, increased competition among stations, and reduced per-station revenues, forced stations to look for low-cost program forms. The low-cost form available was the platter program. As a result, aside from retention of news broadcasts and talk programs, virtually all local programming was platter-music type by 1951–52. The great success of the network telephone quiz show "Stop the Music" in 1948–49 resulted in a number of local telephone quiz programs. Similar to the success of the network-organized "Major Bowes Amateur Hour" in 1935, a flood of locally produced amateur contest programs appeared, but most local telephone quiz shows disappeared before 1951–52.

Television Programs

With television fully underway in 1948, revenues were low and production costs high. A primary consideration was the discovery of low-cost programs. Networks and stations experimented with use of platter music, concentrating chiefly upon direct disc-jockey programs, and in some instances, pantomime by live entertainers to music played on records. Programs of live music were also widely used. Old motion picture files, westerns specifically, were added, along with remote pickups of network sports events. Both program types were low-cost and popular. Also, wrestling and roller derbies came into high prominence. And in the field of children's programs, puppet shows were presented by practically all stations and networks.

Network Television Programs

Programs used on networks were those previously existing on network radio. In fact, many programs were moved bodily from radio to television, or retained on radio, while a television counterpart was carried on the television networks. During the first year of television network operation, a very high proportion of evening time was used for broadcast of sports events. Two variety programs stood out: "The Milton Berle Show," sixty minutes in length, with a production budget of $8,000 a week; and "The Ed Sullivan Show," costing only $5,000 a week. Success of these programs brought great expansion in vaudeville variety programs. In 1950–51, twenty-nine evening variety programs were carried on television networks. Unfortunately, the supply of available acts ran out, and in 1951–52, the number of variety programs was cut in half. Thriller drama programs caught on quickly, although in 1948–49, only two were scheduled for network program development. There were five on the air in 1949–50, twenty-two in 1950–51, and thirty-three in 1951–52.

Daytime network programs followed the pattern set by radio. In 1951–52, 60 percent of all daytime programming consisted of low-cost variety programs. During evening hours, in 1951–52, thirty-six hours a week were devoted to dramatic programs with more than one-third of those shows consisting of the thirty-minute dramatic offering. Totally, fifteen hours went to variety; ten hours to music; fourteen hours to quiz and audience participation; twelve hours to news, forums, and informative talks; fifteen hours to children's programs along with sports broadcasts and miscellaneous forms. Distribution of program types in 1951–52 greatly resembled distribution used on radio networks years earlier and had taken most of the forms in use then.

New Television Network Programs Types

Ad-lib courtroom dramas or reenactments—('48–49) "The Black Robe"
Children's puppet programs—('48–49) "Kukla, Fran, and Ollie;" "Howdy Doody"
Televised sports events—('48–49) boxing, wrestling, roller derby
Western, silent films with narrator—('48–49) "The Hitching Post"
Theatrical feature films—('50–51) "Hollywood Premier Theater"
Actuality demonstrations—('50–51) "Zoo Parade"
Pantomime-to-records—('51–52) Paul Dixon

Local Television Programs

During the years preceding 1950, television stations limited broadcasting operations to evening hours. Those on the air during the daytime, normally went on the air at noon, or later. Most stations experimented with various types of platter music, with little success. Daytime programs included cooking and recipe programs broadcast daily, while the majority of stations presented an hour of audience-participation programs during daytime hours or at night. All stations presented at least one local news broadcast every day, with weather programs, and sports news added to those programs.

Most stations were making extensive use of syndicated film programs by the end of the period. A number of syndicated films for television series, mostly westerns, were available, and during 1950, a large number of theatrical feature films were released for syndication. By the winter of 1951–52, three hundred such features had been released and were included in station schedules.

Seventh Period: 1952–65

This period is highlighted with television gaining the dominant broadcast entity position. It is the period where radio networks were of minor importance in broadcasting. For the most part, trends already established by 1951–52 continued through 1965.

Radio and Television Homes

In spite of the dominant position of television, purchase of radio sets continued, especially portable and transistor models. Through a relaxed import/export trade agreement and the Marshall Plan, Japan and Germany began to find a most receptive American buying public. An estimated twelve million receiving sets for home use were manufactured and sold during 1959; an additional five million automobile sets were manufactured. Since 1957, 98 percent of all homes were radio-equipped, and forty million automobiles came off the assembly lines with radio receiving sets factory-installed. The number of television homes similarly increased to forty-five million homes in January 1960, or 88 percent of all American homes.

Radio Stations

In spite of the increased importance for television, new radio stations were built each year. As of January 1960, thirty-five hundred AM radio stations and seven hundred FM stations were on the air. Of the total number of FM stations, all but fifty commercial stations duplicated AM station programming. Of the thirty-five hundred AM stations, one hundred operated with power of 50,000 watts, while four hundred and fifty operated with 5,000 watts. One-third of all AM stations in January, 1960, were daytime-only stations.

In spite of continued increases in the number of stations, one-third of all AM stations were operating at a loss over the period of 1950 to 1959.

Television Stations

Following the lifting of the freeze on new station authorization in 1952, more than four hundred and fifty new television stations came on the air. As of January, 1960, five hundred and sixty-five television stations were in operation, including ninety operating on UHF channels created in the 1952 order by the Federal Communications Commission. Like radio in previous periods, one-third of all television stations, including all UHF stations and a substantial proportion of postfreeze VHF stations, operated at a loss as of 1958–59.

Radio Networks

The big four radio networks still existed. Revenue from sale of time decreased steadily, from an all-time high of $114 million in 1948 to an estimated $40 million in 1959. Many one-time affiliates turned in affiliation contracts when they reached expiration dates. During 1959, all four networks revised the basis of payment for affiliation stations' time until, by early 1960, affiliates were receiving no cash payments whatsoever from the networks. In a majority of cases, all received only a program service, with major emphasis in actuality and headline news.

Television Networks

The fourth television network, DuMont, ceased operations in 1954, with very few cities having more than three VHF television stations. There simply was not room for more than three networks. The three that survived earned total revenue of $450 million during 1959. CBS continued as the leading network in program popularity and annual revenue, NBC was second, and ABC third, in that order.

During the 1958–59 and 1959–60 seasons, ABC gained equality with the other two, from the standpoint of popularity of evening programs broadcast.

Radio and Television Advertising

Total revenue of radio from sale of advertising in 1958 demonstrated that $540 million was the highest total in history. However, only $45 million went to networks; $172 million represented network spot advertising; and $324 million came from national spot advertising; while local advertising totaled $181 million.

Radio Programming

During the entire period, revenue of radio networks declined. With decreases in revenue and fewer sponsored programs, the networks' program service declined accordingly. Networks experimented with extensive schedules of recorded music, provided thriller dramas, combined established comedians as Amos 'n' Andy, Fibber McGee, and Bob Hope, with recorded music, while continuing to provide news programs. Until 1959, NBC and CBS provided a constantly decreasing number of daytime serial dramas for listeners. NBC discontinued its special serial drama offerings in 1959. No new network program forms were developed, with the exception of the magazine format used in NBC's "Monitor," using a mixture of platter music with short talk spots ranging from comedy monologues to news.

Local radio continued patterns well established by 1951–52, with local platter music programming. With the decline of networks, low-cost programming was essential; therefore, the majority of radio stations went to a news and music format, playing recordings of top-forty popular tunes with a minimum of other types. News changed in character, with most stations providing five-minute news programs. Local disc-jockey programs were modified in the direction of the magazine concept, with short feature items inserted in those platter shows. Only a minority of stations attempted to provide any degree of variety in programming. All advertising on radio stations was the spot-announcement type.

Television Programming

Network television programming was astonishingly similar to types of programs offered on network radio in the period following World War II. The number of comedy variety and straight variety programs were decidedly limited. Half of all evening network time was devoted to various types of dramatic programs. In 1959–60, adult Western dramas were used more than any other. Evening quiz and audience-participation programs disappeared entirely, as a result of the revelation that the big money quiz programs from 1955 through 1958 were rigged and crooked.

There were, however, few musical variety programs, but those headed by Perry Como and Dinah Shore were outstanding. Accordingly, only a single program in concert music aired during this period. Networks limited themselves to a single fifteen-minute news broadcast per day, for each network. A substantial number of public affairs and historical documentary filmed programs were offered. During the daytime, most network program hours were devoted to low-cost variety and audience-participation shows, which were first developed on radio. The remaining time was used to present daytime serials, teenage musical variety programs, and children's programs. Sports events were also featured on weekend afternoons.

Important program types on network radio changed, but those changes came ever so slowly. In the case of network television, changes took place with great rapidity. There were years where emphasis was placed upon broadcasts of sports events on a regular basis. With the fading of sports, variety and vaudeville variety moved into those positions. In 1950–51, networks provided listeners with a number of Western and science fiction thriller programs each week. Three years later, only two such programs were offered in evening hours. Detective drama reached a peak in 1951 and 1952, with twenty-one hours a week devoted to such programs. In the following years, the successful detective shows disappeared. In the 1953–54 period, twenty-eight network hours each week were devoted to prestige drama, and in 1959–60, regularly scheduled prestige drama programs ran an hour or longer with two to three each week. In the fall, 1955, the only big program was "The 64,000 Question." During that season and the one following, six big money quiz programs were on the air. Today, the form has disappeared. More recently, the adult Western reached a peak of popularity; now this type, which remained strong through March 1964, declined in favor of space-adventure, and space-cartoon series, as well as the evening adult serial drama.

Next came jazz detective programs that were highly important, but they began to fade during the 1959–60 season. During the 1958–59 seasons, television networks renewed their interest in sports with telecasts of boxing, bowling, golf, horse racing, basketball, and football on a regularly scheduled weekly basis. During those seasons, where particular sports were important and successful, broadcast "teams" of announcers kept this type moving. Sports Director Roone Arledge, ABC, formulized the successful team of Frank Gifford, Howard Cosell, and Don Meredith in the next period for professional NFL football on a weeknight, which was unheard of in previous seasons. ABC's "Monday Nite Football" moved the popular sports reporting syndrome past the era of bland sports reporting, with a variety of advocacy angles, together with a "love-hate" relationship for Howell Cosell.

One aspect of network television deserves special attention. In 1953, RCA's color television system was officially approved by the Federal Communications Commission. Since 1954, an increasing number of network programs, especially those on NBC, owned by RCA, began broadcasts in color. Color was expensive, so a large proportion of color programs took the form of spectaculars and specials, with some scheduled on a regular basis.

Local television programming, or the nonnetwork variety, included little live local production. The typical television station presented one or two daily news programs, a weather forecast, and possibly a sports news program, with one or two children's programs. The remaining nonnetwork hours were devoted to a syndicated form library, with an average of two showings per day of theatrical feature films, along with two hours each day of syndicated made-for-television thirty-minute programs.

New Radio and Television Forms

Radio—adult western drama—('52–53) "Gunsmoke"
Radio—magazine-type variety—('55–56) "Monitor" (NBC), "Sunday Afternoon" (CBS)
Television—jazz-setting detective ('57–58) "Peter Gunn," "M-Squad"

Eighth Period: 1965–Present

This period is highlighted by three major FCC actions that had a far-sweeping effect on programming for television networks. Also, the rise of large cable television holdings began to develop in the 1970s. This development brought back a large number of previously successful program forms through decisive syndication agreements. Trends established since 1965 continued in television production, while network radio contracted its holdings in the magazine format of news, with a renewed emphasis in sports.

Satellite communications began July 10, 1962, when the National Aeronautics and Space Administration, in cooperation with American Telephone and Telegraph, launched the Telstar Satellite, amplifying signals from this country with ten million times the amplification, and sent those signals back to earth to receiving antennas in France and England.

In other research endeavors, prior to the beginning of this vital period, and the close of the previous one, Hughes Aircraft Company launched a synchronous orbit satellite, reaching a point in space where it rotated at a speed synchronized with the speed of the earth's rotation. This particular Hughes Aircraft satellite, Syncom II, launched in July 1963, was a success of tantamount importance. It marked a milestone, because from that time on, it was not necessary to interrupt programming or any other form of communication through the broadcast sphere when the satellite had left the range of the earth station's signals. Live coverage could now take place and hours of uninterrupted live programming became possible and very popular.

Prior to this remarkable achievement, Congress, in 1962, authorized the Communications Satellite Corporation Act (discussed in other portions of this book) which became the pathfinder for international satellite systems. In 1974, it became the International Telecommunications Satellite Organization (INTELSAT), with a membership of more than eighty nations and presided over by a secretary general. As a result, 95 percent of the world's international communications traffic now comes from INTELSAT, with more than one hundred members today.

In other endeavors, "Early Bird," the first INTELSAR satellite to be launched into orbit, was followed by a long series of INTELSAT spacecraft, orbiting the earth and providing world-wide systems of communication, in computer information systems as well as broadcasting. INTELSAT Series I through V have become the most sophisticated communication satellite programs developed by man.

Numerous domestic satellite systems, beginning with RCA's first system, which originally leased circuits from Canada's ANIK II satellite, have been launched into space. RCA launched the SATCOM I in 1975. Beginning service in 1976, the NBC parent company of RCA launched a second satellite, named SATCOM II, in March 1975 that now has become part of the total RCA communications group network. Meanwhile, Western Union employed a domestic satellite system as well. This satellite is capable of carrying varieties of information specifically to receiving stational systems in New York, Atlanta, Chicago, Dallas, Los Angeles, San Francisco, and Washington.

Radio and Television Homes

In the dominant position of control, television has continued to develop with a new, viable fourth network appearing, in the form of cablevision. One-half of all communities across the nation have approved cable companies for multichannel service and access, including national cable access channels as Home Box Office, CineMax, and a variety of sports network channels including USA, ESPN, and two major entertainment channels—one in Chicago and the other based in Atlanta, Georgia.

As of January 1980, approximately six thousand AM radio stations and three thousand FM stations were on the air. A 1967–68 ruling by the FCC forced AM outlets to cease simulcasting FM through AM outlets. The FCC offered the rationale of minority group access to broadcasting as a realistic principle for part of that major decision. As in the past, one-third of all AM outlets were daytime-only stations.

For years the FCC enforced rules prohibiting after-dark broadcasting by certain small radio stations, while many listeners have had radios go blank as early as 5 P.M. in the wintertime. The Daytime Broadcasters Association, an organization representing more than twenty-three hundred stations limited to daytime broadcasting, has been seeking to lift this restriction, claiming there are six hundred communities and an estimated forty-six million people who are denied nighttime service from a local AM station. The Clear Channel Broadcasters Association denied this, and their views will be discussed later in this book.

However, the association for daytime owners has made some headway to extend operating hours with a measure that went before Congress in the 1983–84 season. A measure by Senator Larry Pressler, Republican, South Dakota, would direct the FCC to allow those stations to sign on the air two hours before sunrise and remain on the air for two hours after sunset. The only reason such authority could be denied under current legislation would be if the daytime-only station caused objectionable interference to the primary signal of any existing station, or would violate international treaties causing interference to stations in Canada or Mexico.

The measure, in the future, is expected to face a very hostile reception on the U.S. Senate floor, because it would trim coverage protection now afforded some of the nation's largest AM radio stations[3] which must shut down at sunset and wait until sunrise to come back on the air. According to the FCC, roughly half of all AM stations in the country operate under such restrictions.

The broadcast restrictions stem primarily from the physical ability of AM stations to produce what are known as skywave signals at night. (This will be discussed in a later chapter.) Historically, in the early days of broadcasting, the FCC purposely licensed a small number of high-powered stations to make skywave signals available at night. The FCC reasoned that those particular signals were the only way rural America could receive radio service.

In this period, an estimated twenty-four million receiving sets for radio units have been manufactured; the nation reaached a total saturation level in access to radio sets; and 100 percent saturation appears likely in the 1980s with television and cable. There is a 46 percent increase in families having not only multiple television sets, but a pay-television function as well, in the top two hundred major markets with cable television hookups.

Radio Stations

The FCC indicated it plans to give preferences to members of minorities bidding for new stations, even when the community around that particular station has no sizable minority population. Without viable options in band space, and with the number of AM and FM outlets on the air, there seems to be no other alternative, without widening the band or closing the spread between stations on the band. An additional option, opposed by the Clear Channel Broadcasters' Association,[4] includes lessening the 50,000-watt AM I-A clear power that these stations currently enjoy.

The FCC action emphasized a policy of increasing ownership of broadcast facilities by blacks, Hispanics, and other minorities, and not of simply providing minority-designed programming and programs by white-ownership functionaries. If the Federal Communications Commission action holds in the 1980s, the Federal Communications Commission will expand those policies to promote more diversity in management of AM and FM radio stations. The Commission, in recent years of this period, followed a policy of giving preference to members of minority groups in competition with others for the right to open new stations or take over operations of existing stations.

Television Stations

As of January 1980, the FCC initiated the "Drop-in" Rule, allowing specifically for minority groups to gain control of small television entities between two large, high-powered stations on the same channel. The smaller, minority-owned stations would be "dropped-in" between signal contours of larger television entities. This would enhance the minority group-control factor, according to current FCC policies and procedures.

Also, cable television markets have opened up new possibilities for a variety of minority communities across America, not only in major television markets. Cablevision, regulated by the Federal Communications Commission, while authorized by local governments, has the public responsibility to contract with cable companies and thereby make a decision from all competing cable entities as to which one will provide superior service. It is the instrument of local government that awards cable facilities to the best of those competing outlets. The outlets themselves must provide multichanneled community access, meeting public interest, convenience, and necessity standards.

In other television-related developments, Communication Satellite Corporation has emerged as a leader in this cable facilities market. The corporation is engaged primarily in the business of providing communications services through facilities of various international, domestic, and maritime satellite systems.

This communications business, as with others in the field, comprises three segments; jurisdictional satellite system services, nonjurisdictional satellite systems services, and communications products and information services. It is this country's participant in the International Telecommunications Satellite Organization (INTELSAT), and the International Maritime Satellite Organization (INMARSAT).

At the time of this writing, through its subsidiary, Satellite Television Corporation, COMSAT is preparing to offer the American family three channels of premium pay television, without advertising, beamed by satellite to individual homes. These satellites will also have the capability of exploring the potential of high definition television. The FCC is now considering STC's application for such a direct broadcasting satellite system, along with applications filed by other companies following their lead. With an FCC go-ahead in the near future, COMSAT would be in a position to undertake satellite construction and initiate satellite-to-home service in eastern parts of the United States by 1986. STC issued a request for bids on the manufacture of satellites to be used in its system. Construction and preoperational costs for the initial phase for their services are estimated to be about $400 million.

As a forerunner to full-scale offerings, STC announced in 1981 it was exploring various alternatives for entering the marketplace in a more limited way before a projected 1986 start-up. STC has investigated opportunities to offer a narrowly targeted pay-television service aimed at a limited attractive customer base in the northeastern parts of the country.

COMSAT is also playing a pivotal role in setting the stage for the commercial introduction of satellite-to-home television. They came forward with the first direct broadcasting proposal in the U.S.

Radio Networks

Several regional sports and special events networks have been created in the past decade to program and target audiences for several local stations. NBC, CBS, ABC, and Mutual still provide a variety of network services to stations in large and small markets. Work has been concentrated upon a revised basis of payment for affiliated stations, with a renewed emphasis upon the magazine format for news. Networks continue to offer actuality and headline news for a variety of stations. Small stations, unable to finance news operations, still rely in large part on the network service. If small stations do not provide network news on the hour and half-hour, at least one of those time slots will be devoted to network radio news. Larger stations rely on networks for news in late-evening and early-morning slots for either the hour or half-hour slot, while circumventing one of those slots for magazine formats of news by the station's own news or public affairs department.

Television Networks

The newest innovation to affect the major television networks is the advent of cable television. The Turner Broadcasting System, based in Atlanta, Georgia, challenged all three major television networks for programming in areas of news, major feature films, and sports. A non-New York–based entity in network programming and functions, Turner provides quality service to a number of cable systems throughout the country.

The Turner Broadcasting System has an in-depth cable news channel function and a headline news service running twenty-four hours a day, seven days a week, over WTBS, the base-cable station in Atlanta, providing a variety of news and entertainment functions for a variety of cable operations. The financial figures over an extended length of time are not yet available on how this will totally affect the three major television networks; nevertheless, it will have a most significant impact.

Aside from cable successes, two most significant pieces of FCC regulatory material have had a most profound effect on network television and station programming. First of all, the 50-50 Rule prohibits networks from sharing in ownership or profits of more than half the entertainment programs in evening schedules, and from engaging in any domestic syndication or foreign distribution of programs they themselves did not produce.

The original staff consideration proposal, headed by Ashbrook P. Bryant, required 50 percent of each network's evening entertainment programming to be supplied and controlled by advertisers. In acquiring programs, networks were limited to the right to broadcast, without any rights or financial interests, even though they finance production of the programs.

Networks, under the plan, would have been barred from domestication syndication and foreign distribution and confined to their own productions. This would have limited networks to ownership or financial participation in 50 percent of the programs in their 6:00 to 11:00 P.M. schedules. It was exclusive of news and sustaining programs. It was modified to specify that networks be allowed to schedule from 6:00 to 11:00 P.M. as much as two hours of nonnews programs, in which they hold equity.

There were three basic modifications in the original plan. First of all, there was a modification made to meet networks' objections, that they would not be able to make plans for their half of the schedule until sponsors and independent producers make commitments for the other half. Secondly, the 50 percent rule was based upon the networks' practice of programming four hours between 6:00 and 11:00 P.M. Thirdly, networks would retain control over the entire schedule and would be able to accept or reject sponsors' programs.

The FCC views this rule as simply limiting networks to equity holdings in no more than 50 percent of nonnews programming between 6:00 P.M. and 11:00 P.M., or a weekly total of fourteen hours, or two hours each night. It also prohibits networks from engaging in domestic syndication and bars them from foreign distribution of independently produced programs. They are prohibited from participating in profit-sharing or distribution rights in such programs as well. It does not lessen networks' responsibility to choose, schedule, and supervise programs they do carry. Finally, it does not affect station licensees' ultimate responsibility for what their particular stations carry.

The second piece of legislation was the Prime Time Access Rule, which went into effect in the 1971–72 season. The three-hour rule on network programming, and the ban of the off-network programs and feature films have had a most profound effect on television stations. The FCC expressed dissatisfaction with past network programming, contrary to program diversity. The rule specifically curtails a half-hour of network programming each night, amounting to 25 percent of the network's prime time.

The FCC observed the rule as a limit to three hours, the amount of network time a television licensee, originally in the top fifty markets, may carry between 7:00 and 11:00 P.M., New York time. It ruled out feature films and off-network programs in the vacated network periods. The FCC said that networks' tremendous domination of programming makes development of other sources desirable, and that there was nothing novel about putting restrictions on licensees, which has been practiced with precedents going back to chain-broadcasting regulations of 1943, as discussed earlier. The adoption of this rule does not violate the Administrative Procedures Act, as well as contentions that the rule required stations to take syndicated programming for periods vacated by the networks.

On the other hand, the rule encourages stations to develop programming for these periods. If smaller stations are harmed, it is not the fault of the rule, but the fault of the networks in refusing to program for markets below the top fifty. The FCC pointed out, additionally, that network domination has been increasing and that access limitations fall within the FCC's obligation to provide for larger and more effective use of broadcasting. Furthermore, the FCC said, at the time, that it did not violate First Amendment rights but, rather, it furthers those aims. Finally, the rule does not interfere with stations' ability to counter programs against competition by using off-network programs and feature films.

Radio and Television Advertising

Independent broadcasters fared rather well in the decade of 1970s. However, ownership shares of network common stock on the major stock exchanges had a lackluster performance, because recent turnarounds of the market of network time lacked strength for the latter part of the decade. CBS, throughout the period, had the best price gains in prime time, but its other operations led investors to lower estimates of per-share earnings and to change intermediate-term ratings on common shares of CBS, to NBC and ABC, for long-term advertising gains.

The network advertising outlook was tagged with uncertainty. The fundamentals remained strong for television stations and cable systems, so there was a positive posture. However, forecasts for the network market of programs and economy continued to indicate a cautious optimism. As was the case in the last period, advance sales of available network time had been relatively low for new seasons in this period. The result was limited predictability for network revenue and profits, and the up-front realization of prime-time prices increased by a moderate 9 to 12 percent, up to 15 to 19 percent for CBS, but up only 6 to 8 percent for NBC and ABC. Negotiating power seemed to rest mostly with buyers of network time throughout the decade, with advertisers increased budget flexibility for buying network time in some cases.

Networks had to reestablish stability programming and audience delivery for the latter half of this period because of a loss of status. This was due, in part, to cable-television penetration in major markets and FCC rulings on the 50 percent and prime time rulings. Prices for daytime and other daytime programming were, on balance, less strong than prime time counterparts on all networks. Revenue gains for the latter half of the 1979–80 season widened in comparison with particularly weakened, earlier yearly levels. Total revenue increased by 9 to 10 percent in 1981–82, with a gain of 13 to 14 percent for CBS. The gain in total revenue for 1982 was projected, at that date, at 9 to 12 percent, with an increase of 12 to 14 percent for CBS, 9 to 11 percent for ABC, and 7 to 9 percent for NBC.

National "spot advertising" revenue remained strong, increasing at a rate of about 14 percent; however, there was some improvement in local spots beginning in the latter half of the decade, with a gain of 9 percent in the 1979–80 season. Stations benefited from flexibility and buyer disenchantment with networks, while independents continued to outperform the industry. The gains in 1981–82 season increased 11 to 12 percent in local revenue, and from 12 to 13 percent in total station revenue in 1981–82. Total station revenue rose by 11 to 14 percent in that 1981–82 season.

Summer ratings duplicated patterns of past regular seasons, with CBS leading in prime time, ABC in daytime, and NBC a weak third in both. No changes occurred for the new season, despite NBC's last-minute prime time changes. ABC evening news outrated CBS for one week in July, but Dan Rather, a noted CBS news commentator, regained the lead in subsequent weeks. There were closer races in news in the new season.

April subscriber counts in 1981–82 cable advertising showed that MSO Cable grew faster than the industry. ATC Cable took over the lead in total subscribers and retained it in pay units, up to 70 percent in a year over a million. Storer Cable pay units were 179 percent; the Times-Mirror Media Group showed 97 percent; Cox and UA-Columbia were up more than 70 percent; and Telecommunications was up to 66 percent. The pay-unit penetration of basic subscribers rose to 60 percent, due to better acceptance in new builds and better overall marketing for cable across the country.

Notes

1–4. Please check the station ownership record in another section of this book.
5. Educational television and related programming will be discussed in an additional section.

Chapter 4

The Broadcast Industry in the U.S.

Broadcasting and cable are not simply matters of station and network operations. The industry includes a wide variety of other types of business enterprises and concerns, dependent upon and connected with broadcasting and cable. Most phases of the industry were developed in pretelevision days, but with the rise of television and cablevision, functions already performed for radio were extended to meet requirements for television and cablevision, or in some instances, new types of companies came into existence to provide the services needed. The major components of the industry should be enumerated.

RADIO STATIONS

As of January 1980, a total of over six thousand AM, or standard-band radio, stations were on the air in this country, and approximately three thousand FM stations were also functioning. These totals also include AM and FM stations operating as educational broadcasting stations, or those functioning on a noncommercial basis. Of the commercial FM stations, all provide a separate program service, and are not duplicated with AM counterparts. The one exception to this rule includes minority-owned stations, where the FCC makes provisions for those specific operations. Examples include the Oglala Sioux Broadcasting Corporation, and Navaho National Broadcasting Corporation, Window Rock, Arizona. The Oglala Sioux Broadcasting entity duplicates AM broadcasting with FM operations, as well as UHF and VHF television functions, from their area of operations in South Dakota. This also applies to Chicano and Native American endeavors elsewhere.

Of approximately six thousand AM stations, more than twenty-five hundred came into being after the close of World War II, and in 1945, only 950 AM outlets were on the air. Similarly, nearly all FM stations were first authorized after the close of the war, and fewer than twelve were on the air at the end of 1945.

Of the total number of standard-band AM stations, one hundred operate with 50-kw power and are classed as "Class 1" stations. Others are authorized to use lower power. Two-thirds of all AM stations are licensed to operate only during daytime hours, going off the air at night to avoid interfering with signals of Class 1, 50-kw stations. Of commercial AM stations on the air in January 1980, over one hundred and fifty are owned by groups owning more than one AM station, with over six hundred multiple-station owning groups. The remaining commercial AM stations are independently owned, with one station owner or individual having no substantial financial interest in any other AM radio station.

TELEVISION STATIONS

As of January 1980, there were over four thousand television stations on the air on the VHF band, including a number of noncommercial educational stations, and 1120 UHF stations. This also includes a number of noncommercial educational stations. Practically all stations came into existence after the end of World War II. At the end of 1945, only seven operating commercial stations were in existence, and only 108 were on the air as of July 1952.

A majority of commercial television stations originally were owned by individuals or concerns operating AM radio stations in the same communities. The FCC now applies "concentration of media" rules to those outlets in order to allow minority groups increased access to broadcasting and cable facilities. A number of television stations are owned by groups owning two or more television stations, and there are more than four hundred such multiple-television owning groups.

Originally, seven of the commercial stations, six VHF and one UHF, operated as satellite stations, originating no local programming, and simply reproducing signals of individual originating stations in a different locality. Similarly, a number of regular television stations, including educational television stations, feed their programs to translator stations in other communities. A translator station is a low-powered

UHF station, originating no programs, that picks up signals of a regular television station, converting the signal to another frequency or channel, always UHF, and rebroadcasting it.

CABLE TELEVISION

Also providing television programming via underground cable facilities to television equipped homes in autumn, 1980, were several hundred community antenna systems. The operator of a community antenna system simply erects a very tall tower, or satellite "dish" to pick up signals of several television stations, or access channels, from a communicational satellite in a fixed orbit above the earth. This satellite system operates through a system of cable facilities, microwave relay, and direct satellite communication reception to receiver sets in a variety of homes connected by coaxial cable for a rental fee per home served. Community antenna systems are not broadcasting stations in the strict sense of the term. However, the FCC adopted rules and regulations related to cable facilities, owners, and outlets.

At the outset of CATV systems, there was a legitimate argument that cable systems were not broadcasting stations because they provided no signal through the air, and therefore were not subject to licensing as standard broadcasting stations. It was for this reason the FCC adopted specific rules and regulations that govern these systems through a local government entity.

RADIO NETWORKS

Numerous radio networks operate now on a national basis. NBC started operations in 1926, CBS and ABC in 1927, with ABC as a wholly owned entity of NBC, while Mutual Broadcasting System was created in 1934. As of January 1960, NBC and CBS had over two hundred affiliated stations. Now, other networks offer a variety of regional news and programming material for stations. These networks target news and programming material for specific radio age groups and audiences, while the selectivity factor includes specific stations in the AM or FM variety.

Prior to the rise of television, radio networks were the dominating force in the broadcast field. Since the advent of television and cable dominance, most national advertisers, who formerly provided programs on radio networks, switched to the new entities. The result has been radio network revenue shifting to specific targeted audiences with varieties of musical tastes and demands.

In addition, there are several regional networks operating on an informal basis, existing only to carry broadcasts of collegiate or professional sports. As with national networks, stations included in these regional networks rarely receive cash payment for carrying regional sports programs. They are allowed to carry the sports programs for set fees, and insert spot announcements for local advertisers, according to the sports, university, or professional sports affiliation agreement. Collegiate sports broadcasts are also controlled by the CFA, NCAA, NAIA, and NJCAA, while professional sports are controlled by professional associations connected with those areas of competitive athletics.

TELEVISION NETWORKS

Television networks were first organized in 1948. Originally, there were four network companies, ABC, NBC, CBS, and DuMont. After a few years, DuMont discontinued its network efforts, leaving three television networks operating on a national basis. Stations affiliated with one or more of these networks serve markets with coverage involving greater transmission costs than advertisers are willing to pay locally. In markets with three networks, each network has a primary affiliation arrangement with one of the available stations. In two-station markets, the ordinary practice is for one station to be affiliated with two networks, depending upon programming requirements.

In one-station markets, that station will have affiliation arrangments with at least two of the three networks, and often with all three. The basis of network payment is similar to that used in radio network affiliation agreements, prior to the rise of television. A station with primary affiliation with a network is paid an amount ranging from 25 percent to 35 percent of its rate card, on a national rate, for the hours used by that network for commercial programs. The station's rate for daytime hours is normally half the amount of the evening rate. In addition, stations give networks an option of certain hours specified in individual contracts. This applies for broadcasting time every day. Originally, it was those three hours in the period between 8:00 A.M. and 1:00 P.M., together with an additional three hours between 1:00 and 6:00 P.M., and a final three hours between 7:00 and 10:00 P.M., called prime time.

In exchange, the station has the benefit of carrying network commercial programs and providing a stipulated number of hours each week of sustaining programs that it may or may not use. The network programs are generally stronger audience receivers than those programs presented locally by affiliates.

AMERICAN TELEPHONE AND TELEGRAPH COMPANY

Until recently, AT&T had a complete monopoly in the broadcast industry. It held monopoly power in transmission by coaxial cable, microwave relay, ordinary telephone lines, as well as radio and television programs between cities. With cable, the Communications Satellite Corporation has virtual control of satellite technology in multichannel facilities for cable outlets. In few localities, where AT&T has not provided service, the carrying of programs may be handled by local concerns, but 99 percent of all program transmission by wire, or microwave relay, has been with AT&T. The FCC in 1982 ordered the Bell System to divest itself of this particular company, along with others that constituted a monopoly.

For national networks, radio, television, and cable to operate, the services provided by AT&T and COMSAT are essential. A great deal of money is paid by the three national networks, as well as cable networks and operators, for transmission of programs. However, the total payments for transmission of satellite communications vary from year to year.

EQUIPMENT MANUFACTURERS

Manufacturers of radio and television equipment have total revenue from sale of those products in excess of the total annual revenue of all networks and stations from sale of time to advertisers. Such companies as RCA, Tandy Corporation, General Electric, Westinghouse, and COMSAT provide equipment to the industry. Many companies provide antenna towers, tape recorders, integrated circuitry and circuit boards, switching systems, production units, control room equipment, studio lights, film projectors, cameras, and a variety of film, video cassettes, and tape-editing equipment.

Minnesota Mining and Manufacturing Company, or 3M, and Eastman Kodak produce most of the magnetic tape used for audio and video tape recording. Kodak also supplies a large amount of film products, while TelePromTer Corporation provides electronic cueboards to stations and networks. The Jerrold Corporation, Electronics, specializes in community antenna systems, and Kliegl Brothers produces motion picture and television studio lighting equipment. A variety of foreign manufacturers, such as Sony and Mitsubishi, specialize in camera lenses, with self-focusing television cameras and "zoom" lenses.

NATIONAL ADVERTISERS

Without national advertisers, network and cable television would be impossible. They buy network time, which is very high. The average evening television network program, sixty minutes in length, will cost an advertiser $500,000 in program costs, and another $100,000 to $200,000 for network time. Television specials cost as much as $500,000 to $1,000,000 for production, with an additional $500,000 for network time. In addition to buying network time, national advertisers buy program time, or time for commercial spot announcements on local stations directly.

National advertisers spending money on television network advertising include Proctor & Gamble, General Foods, American Home Products, General Mills, General Motors, Pepsico, Lever Brothers, Ford Motor Company, AT&T, McDonald's, Westinghouse, RCA, Gilette, Budweiser, Coors, Ernest and Julio Gallo Wineries, Chrysler Corporation, and many others.

On the local level, there are literally tens of thousands of advertisers. A few buy time for local programs on individual radio or television outlets, or on cable-access channels. Programs include news, sports, weather, syndicated programs, and specials. But the majority of local advertising money is spent for spot announcements inserted in radio or television programs, during station and network breaks. Local advertising accounts for nearly two-thirds of total revenue of radio stations in the United States, taken collectively. In smaller communities, all revenue of radio stations comes from local advertisers.

ADVERTISING AGENCIES

All network and cable, all national spot advertising on broadcast stations and cable outlets, and 30 percent of all local advertising are placed by advertising agencies. An advertising agency is a company specializing in advertising for other concerns. When an agency is employed by a national advertiser, the agency plans the advertising campaign, including print, direct mail, billboard, as well as advertising on radio, cable, and television. The advertising agency selects the program to be sponsored where that advertiser's announcements will be inserted; they then write the copy and produce or arrange for production of live, filmed, taped, or "still" shot commercials, while handling all program promotion, including the care of detailed arrangements with broadcast or cable entities.

In case of national nonnetwork spot advertising, the agency plans the campaign, writes the copy, prepares commercials to be used on film or tape, selects the station or stations to be used in each market, and then contracts with stations selected for carrying those announcements. For the services they render, advertising agencies, receive a commission of 15 percent of the gross cost of the network, cable, or station time purchased. It is rebated by the broadcast or cable outlet. In addition, they add a commission of 15 percent on the cost of the production of those programs or commercials, paid by the advertiser. In all, there are approximately eight thousand advertising agencies in the United States. Some are extremely small and operate on a local basis. Others specialize in direct mail or billboard advertising, as the case in some small markets where the Don-Rey Media Group functions, as an example.

However, there are many advertising agencies that represent advertisers purchasing radio, cable, or network time, buying spot-announcement advertising on individual stations on either a regional or national basis. On the basis of total volume of radio, television, and cable, the largest advertising agencies for broadcasting include:

1) Young & Rubican
2) J. Walter Thompson
3) Batten, Barton, Durstine, and Osborn
4) The Ted Bates Company
5) Benton and Bowles
6) Leo Burnett Company

7) Dancer, Fitzgerald, and Sample
8) William Esty Company
9) Kenyon & Eckhardt
10) McCann-Erickson

STATION REPRESENTATIVES

Business concerns commonly referred to as "station reps," or simply "reps," have developed and grown with the industry. Since a single station cannot afford to have its own salesmen calling on scores of national advertisers who buy national spot time on individual stations, companies have come into existence that have taken over the selling job for stations, as far as national advertisers are concerned.

These "station rep" concerns have representative contracts with a number of different stations, in different cities throughout the country. In most cases, the "rep" concern has offices in New York, Chicago, Denver, Detroit, Dallas, Atlanta, and on the West Coast, so that there is an office reasonably close to the headquarters of every national advertiser as well as the home office of each of the most important advertising agencies. If an advertising agency is planning an advertising campaign for one of its clients, it is the rep's business to find out about it, and if the campaign involves purchase of spot advertising time on individual stations, they must sell their own lineup of stations, or cable affiliates, to the agency who plans that campaign on behalf of the advertiser. Station representatives also assist stations in planning their own advertising in trade publications, and professional journals.

For its services, the station representative collects a commission of 15 percent of all business it secures for the station, paid by that individual station. About fifty-five rep concerns represent twenty or more stations each, and in some cases, where small stations are represented, the company's list may include as many as sixty to seventy stations. Practically all rep companies specialize in the type of stations represented. For example, nine companies represent commercial stations in other countries such as Mexico, Canada, and Europe. Others represent only fifty-kw stations, or possibly five-kw stations with large areas of coverage. Some rep concerns handle only country music stations, others handle soul stations, while still others handle all talk radio stations. Seven concerns handle only television stations, and there are over twenty that handle cable outlets.

Among the more important rep concerns, based on volume of business handled rather than number of stations represented, are:

1) Peters, Griffin, and Woodward
2) The Katz Agency
3) Edward Petry Company
4) John Blair Company
5) Adam Young, Incorporated
6) Weed Radio Corporation
7) Henry Christal Company

PROGRAM PACKAGE AGENCIES

A "package" agency is a program-producing concern, without broadcasting facilities of its own. It produces programs for use by sponsors, networks, or stations. The package agency owns the rights to the program idea. It plans all programs, employs writers, producers, directors, actors, and technicians, and takes care of all details involved in putting the program on the air. It may also put programs in transcribed, taped, or filmed form, for later airing. It pays all costs of the program, except costs of network or station time. It charges the sponsor or broadcaster a fixed amount, determined in advance, for the entire package, hence the term "package agency."

Package agencies originated in the early days of network radio. One of the earliest packagers of sponsored network programs was Phillips Lord, who created and produced thrillers on radio networks, such as "Gangbusters," "Mr. District Attorney," and "Counterspy." Other important radio packagers included Carlton B. Morse, who produced "One Man's Family," and "I Love a Mystery." Frank and Anne Hummert packaged a dozen different daytime women's serials for radio, while Ralph Edwards was responsible for the game show "Truth or Consequences" and "This is Your Life," produced for network radio.

These package concerns limited efforts to producing live programs for sponsorship on radio networks. However, after 1945, a number of companies began providing programs in transcribed form for individual radio stations under local sponsorship. The most important producer for individual radio stations was Frederick W. Ziv, although there were many others.

Today, with the decline of network radio, network radio packagers have all but disappeared, and while packagers who provided taped or transcribed radio programs for individual stations are far from important, some do exist. It is still possible to buy or rent any of several dramatic program series, daytime serials, or limited numbers of musical programs for radio use. Occasionally, a major radio-station-owning group will produce a documentary series for radio that is also made available to stations other than those owned by that group, on a rental basis. One of the most humorous packages comes from a St. Louis disc jockey who put together a series of short humorous stories about "A Great White-Winged Warrior-Chicken-Man!" This package series ran on a number of major market radio stations during the 1960s, and was most successful in radio packaging.

Packaging, however, has reached its height in television. It is most important now, with the stringent limitations placed upon prime time by the Federal Communications Commission's Prime Time Access Rule, as well as the 50-50 Rule. Some companies specialize in production of filmed programs for individual stations on a rental basis. Among them, Ziv TV, the television counterpart of the Frederick W. Ziv Company, originally active in radio, is an excellent example. Others intended for syndication are Guild Films, Harry S. Goodman, Screen Gems, and certainly Desilu Productions.

Other package companies produce for network television and cable only, specializing in live productions rather than filmed programs. The Mark Goodson–Bill Todman label carries mostly all network quiz programs. A company headed by Lou Cowan was responsible for the development of big money quiz programs. Also, David Susskind's Talent Associates Company produces live dramatic productions for networks.

A third group of television and cable television packagers produce programs on film for immediate network use. If programs aren't sold to sponsors or networks before a series is scheduled to begin, normally no more than one or two episodes are produced. But the idea is to make the film available for syndication or for sale directly to television or cable outlets, for single outlet use, after the network run has been completed. Important packagers of this type include Walt Disney Productions, Warner Brothers/Seven Arts, 20th Century Fox, Four Star Productions, and numerous others. Again, as mentioned previously, during seasons since 1971–72, all evening sponsored network television and cablevision programs were produced by independent package agencies or by package agencies working in partnership and cooperation with networks, due to the 50-50 Rule.

PRODUCERS OF FILMED OR TRANSCRIBED COMMERCIALS

Most commercial announcements used on television and cablevision networks, and most national spot advertising for radio or television are produced by advertising agencies. However, many times these agencies turn to professioanl producers of commercial announcements to provide those commercials if anything special is desired. For example, placing a car on a high mountain, in Oak Creek Canyon, near Sedona, Arizona, or an ocean breeze and shoreline on the Pacific Coast, might be desired for a cosmetic line of commercials. There are concerns which produce filmed television commercials for these special needs and ramifications. Jerry Fairbanks Productions, Sarra, Incorporated, Alexander Film Company, and Academy Film Productions produce commercials planned by advertising agencies and requiring special effects and creative stratas that radio or television production studios cannot provide. In others, particularly, in the case of Sarra, Incorporated, the planning as well as execution of the commercial is handled by the producing company, itself.

A related function is writing commercial jingles or music. Frequently, words used in singing commercials, whether for radio or television, are given to companies employed specifically for this type of production service. The Johnny Mann Singers have produced a number of singing station jingles for stations through the years, including station identification, weather jingles, and record "intros" and "outros."

PROGRAM SYNDICATION COMPANIES

Reference was made to package agencies producing syndicated materials. Programs are made available on a rental basis, from time to time, to individual stations or sponsors, buying time directly on individual stations, rather than buying network time. A syndication company is one engaged in sale of programs to stations, or local sponsors. Sometimes the syndication concern is also a package producer. But more frequently, the syndication company handles sale of programs produced by other companies. The term syndication does not imply that the syndication company is a packager. In a majority of instances, the package agency and syndication company are two entirely different business entities.

Materials syndicated for television, and now cablevision, fall into four major categories. First, there are programs prepared in film or videotape for television and cablevision use. In some cases, these programs were originally presented on networks and are available for syndication under a different name, after the network run has been completed. Frequently, programs are produced for syndication in the first place and are never used on national television or cablevision networks.

A second type of syndicated television or cablevision program material is theatrical feature films originally made for showing in movie theaters. Television and cablevision rights for these pictures have been purchased from companies who originally produced or distributed them for theater showing. These films are rented or sold to television and cable outlets and are shown locally on a participating-sponsorship basis.

A third type of syndicated material includes cartoons and comedy films originally produced for theater showing as well as television and cable options. These materials are handled by syndication concerns who purchased the rights for them. There are few television or cable companies that do not make daily use of one or more of these series, especially animated cartoons which are used over and over again.

Finally, some syndication concerns specialize in educational film or videotape originally used as short subjects for showing on the Public Broadcasting Network, as well as for classroom use, for educational talkback facilities, and for showing on cable access channels. Important among syndication companies in television are ABC, NBC, and CBS films, which are divisions of the network companies that the 50/50 Rule has affected in detail. Others that specialize in feature films include Ziv TV, Screen Gems, Guild Films, NTA, MCA's Film Syndication Division, Associated Artists Productions, Interstate Televevision, and Hollywood Television Service. There are also syndication companies in radio, but this once moderately significant part of radio has now become very minor in importance.

ORGANIZATIONAL, RELIGIOUS, AND EDUCATIONAL PROGRAM SUPPLIERS

Related to syndication companies, but differing in some respects, are organizations engaged in supplying no-charge programs to television, cable, or radio stations. Referred to as program suppliers, they are not syndicators, since syndication companies provide services and programs on rental or purchasing plans. For example, some universities provide educational films used by educational or commercial television stations, without charge, except for fees to cover shipment. Various religious organizations provide programs, sometimes in film form, for television and cablevision, as well as taped or recorded programs for use on radio. Industrial concerns make filmed programs, produced as promotions for their companies. The federal government also provides films in transcribed form for radio stations and film forms for television and cable outlets. These program suppliers might be classed as syndicators, but since their program materials are provided without charge, on a noncommercial basis, they are considered organizational program suppliers.

TRANSCRIBED, TAPED, OR FILM NETWORKS

Related to syndication companies are organizations operating on a limited network basis, using mechanically recorded programs instead of network lines or communication satellites. A pioneer in this field was Keystone Broadcasting System, founded in the 1950s; at its peak, it held contracts with several hundred radio stations, supplying these stations with sponsored transcribed programs. In most cases, they were carried on regular radio networks, but these programs were also transcribed off network lines and provided to stations in small markets not reached by conventional networks. With the decline of network radio, Keystone provided sponsored transcribed programs to radio stations, on a limited basis, prior to 1950.

A second network was the National Association of Educational Broadcasters Taped Network. Inaugurated after World War II, it provided audio taped educational programs to educational stations that were members of the NAEB. In recent years, the same organization provided kinescoped and videotaped programs to members of that national association.

During the 1956–57 season, a national film network organized by National Telefilm Associates, a major syndication concern, took options on time of commercial television stations, supplying each station with theatrical feature films each week and attempting to find national sponsors to pay costs of film service provided, as well as the cost of time on member stations. The experiment was not successful. Sponsors showed little interest in national sponsorship of theatrical feature films. NTA, at the present time, provides several programs in videotaped form, in connection with regular syndication services for television and cable outlets.

SERVICES SUPPLYING PROGRAM MATERIALS

There are companies engaged in supplying materials used in locally produced programs, including news, music, and features specifically.

News Services

The most important news organizations serving radio, television, and cable are the two who provide national news service for the print media as well—the Associated Press and United Press International. UPI is a consolidation of United Press Association and the International News Service, which was owned by the Hearst organization. UPI was recently sold to a group of Tennessee investors. Both UPI and AP provide radio, television, and cable services to the industry. In addition, many broadcast stations buy regular newspaper services. Both companies provide still-photo and news-film service to outlets.

Music Library Services

Stations and cable outlets use music in addition to current popular music found on most disc-jockey radio programs. Music is used for themes, bridges in dramatic programs, and special programs. Several concerns, including Lang-Worth, World, and NBC-Thesaurus provide stations with music libraries of transcribed music, including classical, jazz, rock, blues, soul, country/western, religious, patriotic, folk, and widely varying top 40-hits in separate selections. Libraries are provided on a sale or rental basis.

Sound Effects

Today, several make use of transcribed sound effects, anything from the sound of a train leaving a station and jet planes taking off, to audiences laughing, clapping, or applauding. Sound effects may be purchased separately or in library form. Two of the leading suppliers of transcribed sound effects are Standard Radio Transcription Service and Langlois Film Music. Speedy-Q also provides a service.

Special Effects, Radio

In recent years, many radio stations have made heavy use of special effects such as jingles for station identifications; sound effects to introduce news bulletins and news programs; musical weather reports; and musical productions and closings to talk shows. They are used as needed or transcribed with a variety of promotional "plugs," by name entertainers, on subjects ranging from recipes to thoughts-for-the-day. These special effects are available in ordinary syndication form and are purchased outright, rather than rented. They are also built to order, to meet needs of individual stations.

Special Visual Materials for Cable and Television

Several concerns provide visual materials for television and cable. All are made to order for these outlets. Included are special station identifications and construction of sets. Stations or cable in small markets make their own visuals. All slidemaking or processing of film is handled by station staff members.

Music Licensing Services

Music used on radio, cable, and television, and music written within the last eighty years, is copyrighted and may be performed only after the user has received a license or permit for use from the owner of the copyright. Music licensing services act as representatives for composers, lyricists, publishers, and any others who hold copyrights on music. ASCAP, the American Society of Composers, Authors, and Publishers, controls copyrights on most music and standards from broadway shows and motion pictures, as well as on all popular music. BMI, or Broadcast Music, Incorporated, started by broadcasters, owns a considerable portion of the popular music written since 1960. And SESAC, the Society of European Singers, Artists and Composers, controls rights on most music by European or Latin American composers, as well as some popular country western music of American origin. All American broadcast stations hold blanket licenses covering performance of music controlled by the first two entities. Each station pays approximately 3 to 3½ percent of its total revenue from sale of time, in royalties, to companies representing music copyright holders.

Talent And Artist Agents

Every vital entertainer, and those of minor importance, are represented in business dealings by talent agents or personal representatives. These agents arrange for bookings on a variety of broadcast programs, handle personal publicity, and negotiate with producers or possible employers, all for a 15 percent fee of the entertainer's gross earnings. Although there are dozens of one-man agencies, three highly important concerns that dominate the field are Music Corporation of America, or MCA, the William Morris Agency, and the Jim Halsey Agency. All handle business negotiations and publicity for most entertainers, comedians, actors, masters of ceremonies, vocalists, and orchestras. All engage in the package program business as well as the talent agency field. Quite naturally, program packages they put together feature their own artists. In addition to entertainment, most writers of programs for network, cable, or package programs are also represented by agents. However, these agents are part of a different group that attempts to sell services of writers or to market scripts already written to producers.

Radio And Television Unions

Every aspect of broadcasting and cable is 100 percent unionized. Some networks have union contracts with twenty-five to thirty different unions, covering services of employees of various types. Most stations in cities with populations of 200,000 or more hire union announcers, musicians, and engineers. In smaller cities, few stations have union contracts, although many musicians employed will be members of the Musicians' Union or the American Federation of Musicians. Unions most important in connection with the production of radio, television and cable programs, live, film, or taped include the following.

AFM

The American Federation of Musicians has jurisdiction over those playing musical instruments, handling music libraries, or scoring music in all major cities except Los Angeles. Jurisdiction over musicians in Los Angeles is held by the Musicians' Guild of America.

AFTRA

The American Federation of Television and Radio Artists has jurisdiction over announcers, actors, vocalists, and other entertainers appearing before camera or microphone. Newscasters are exempted in some areas, since they are not always classed as "entertainers."

IBEW

This stands for the International Brotherhood of Electrical Workers.

NABET

The National Association of Broadcast Engineers and Technicians. In some cities, one or the other union has jurisdiction. Boundaries cover announcers, entertainers, and vocalists where there is no adequate representation by AFTRA or AFM. They cover projectionists, cameramen, and technical personnel. NABET has contracts covering engineering employees of all major networks, including cable.

IATSE

The International Association of Theatrical Stage Employees is a widely known union. Originally a union of stagehands in theaters, IATSE also gained jurisdiction over news and motion picture photographers, as well as cameramen and painters. In some cities, scenic designers, lighting technicians, film projectionists, and film editors are covered. In some cities, IATSE covers motion picture cameramen.

WGA

Writers' Guild of America is a union resulting from the consolidation of the ScreenWriters' Guild and Radio and Television Guild of America in 1958. It has contracts with all national television and cable networks. It covers writers of independent package program producers and with major advertising agencies, as well as staff employees and freelance writers.

DGA

Directors' Guild of America was consummated in 1959 with the union of the Radio-TV Directors' Guild and the Screen Directors' Guild. It represents producer-directors, casting directors, or videotaped television or network program directors. It has contracts with national cable and television networks, as well as with most television package agencies. The Guild will not represent directors if they provide filmed programs for companies with television and cable connections.

In addition to these major unions, there are unions of office employees, locals of the Teamsters' Union that have jurisdiction over moving scenery or props from one building to another, and locals for hairdressers, make-up artists, costumers, and designers.

BROADCASTING RESEARCH ORGANIZATION

There are numerous research organizations engaged in radio and television audience research. One group includes the American Research Bureau, which has a subsidiary called Arbitron; C. E. Hooper;

A.C. Nielsen; PULSE; Trendex; and Videodex, which provides regular ratings on a national or local basis. These ratings serve as indexes for popularity and audience size of radio or televison programs.

Secondly, a number of concerns, among them the Institute for Motivational Research, the Psychological Corporation, and Schwerin Research Corporation, specialize in quantitative audience research for the effectiveness of programs or commercial announcements. Some companies monitor national advertisers' commercials carried by stations to make sure stations air those announcements paid for. Accordingly, there are several market research concerns engaging in marketing conditions and the effectiveness of radio and television advertising only.

TRADE PAPERS

Like every other major industry, broadcasting has its own trade papers. Among them are *Broadcasting, Television, and Radio-TV Daily*. Entertainment industry publications such as *Variety* and *Billboard* devote a considerable amount of space to radio and television, as do general advertising trade publications such as *Advertising Age, Printer's Ink, Tide*, and *Sales Management*. There is also one academic publication devoted entirely to broadcasting, *The Journal of Broadcasting*, issued numerous times yearly.

OTHER INTERESTS

Included as part of the industry are trade associations such as the National Association of Broadcasters, Radio Advertising Bureau, Television Advertising Bureau, in addition to state associations of broadcasters and special groups such as Clear Channel Broadcasters' Association, the Daytime Broadcasters' Association, the National Association of Educational Broadcasters, the National Association of FM Broadcasters and the National Association of Educational FM Broadcasters.

There are consultant specialists, including engineers, who assist in preparing applications for new stations or who plan engineering requirements for stations. Broadcasters also make use of program consultants, management consultants, and attorneys who specialize in communications and broadcast law. These individuals assist broadcasters and patrons by appearing before the Federal Communications Commission for station licensing.

Educational groups interested in broadcasting include the NAEB, and the National Association of Educational Television and Radio Center. Also, the Joint Council on Educational Television and the Association of Professional Broadcasting Education are among those best known for improvement of television. In addition, there are many church groups monitoring programs and acting as consumer advocates for what they consider broadcasting and cable's shortcomings.

THE FEDERAL COMMUNICATIONS COMMISSION

Any listing of components of the broadcast and cable industry must include the Federal Communications Commission, the government agency regulating broadcasting. The FCC licenses all broadcasting stations and oversees rules for cable, transmitter apparatus, and engineers employed by stations. The FCC grants or withholds renewal of licenses of stations and exerts a substantial degree of control over operations of stations. It regulates cable, as well as radio and television, through rules and regulations for cable companies' equipment and entities.

Empowered by the Federal Communications Act of 1934, as amended, seven commissioners, chosen by the President and approved by the United States Senate, direct activities and set policy for the Federal Communications Commission. No more than four can be from the same political party. Their duties are administrative, policy-making, and judicial.

The commission delegates specific operating responsibilities to various offices and bureaus. The chairman coordinates and organizes work and represents the FCC in legislative matters and in communication with other government departments and agencies. The executive director coordinates all staff activity of the FCC. He is directly responsible for internal administrative matters. He reports directly to the commission, and works under supervision of the chairman, assisting in carrying out the commission's organizational and administative functions.

Under the general direction of the defense commissioner, the executive director coordinates defense activities of the commission, under the Emergency Communications Division. This includes, but is not only limited to, the National and State Industry Advisory Committees of the Emergency Broadcast System. The general counsel is responsible to the commission on legal matters involved in establishing and implementing policy. He acts in regulatory matters where more than one bureau is involved, as well as in international communications matters. The general counsel coordinates the commission's legislative program and represents the commission in the courts.

The chief engineer advises the commission in engineering matters as they relate to establishment of policy and implementation, and he collaborates with other bureaus in formation of technical requirements of the rules and regulations and advises the commission on such matters. Other operations of the FCC are directed by the secretary of the commission, with respect to development of the Office of Information; Hearing Examiners, who are administrative law judges, including the review board; and Office of Opinions and Review, which assists and advises the commission in reviews and drafting final decisions.

The Broadcast Bureau assists, advises, and makes recommendations to the commission with respect to development of regulatory programs for the broadcast services. Within the Broadcast Bureau are the Office of the Bureau Chief, Broadcast Facilities Division, Renewal and Transfer Division, Hearing Division, Rules and Standards Division, License Division, Research and Education Division, Office of Network Study, and the Complaints and Compliance Division.

The Field Engineering Bureau is responsible for all commission engineering activities performed in the field relating to broadcast stations. It detects violations of regulations, monitors transmissions, inspects stations, investigates complaints of frequency interference, and issues violation notices. The bureau maintains field offices and monitoring stations throughout the United States. It examines and licenses operators, and it processes applications for painting, lighting, and the placement of antenna towers.

Chapter 5

Classification and Definitions

A number of terms used in relation to radio, television, and cable programs and the materials used in programs should be explained. There are a number of general classifications used where advertising is presented with regard to the types of sponsorship of programs. The terms apply equally to radio, television, and cable.

CLASSIFICATION OF PROGRAMS

These various classifications used are self-explanatory. On the basis of frequency of broadcast, the majority of programs may be classed under three headings.

One-time Programs

These are special programs presented on a one-time basis, such as presidential addresses, broadcasts of the Kentucky Derby, spectaculars, or special programs on television and cable networks not scheduled on a regular basis. They include broadcasts by political candidates during campaigns and special news events or sports broadcasts.

Once-a-week Programs

Regular program series are presented on a once-a-week basis, on the same day, at the same hour each week. They include all evening television, as well as daytime and Sunday, on a regularly scheduled basis. This applies to all stations, networks, and cable outlets. Sports broadcasts are classed as once-a-week programs, regardless of the limited length of football seasons.

Across-the-board Programs

These are programs broadcast on regular schedules, five days a week, Monday through Friday. The classification includes all daytime programs, whether network or local, broadcast weekdays. It also includes all Monday-through-Friday late evening programs, network and local. Local broadcasts of major feature films, network broadcasts of late-night, low-budget, variety programs, as well as all early evening television and cable news broadcasts locally produced, or originated by networks, are included.

PRODUCTION FORMS

On the basis of the form in which the program is produced or provided, programs fall into five major categories, one applying to all, two applying only to radio, and two applying only to television and cable.

Live Broadcast Productions

These programs, including delayed broadcasts, in the live category if they are presented as they are produced over all stations.

Programs Produced on Film

These are found on television and cable. Produced weeks or even months in advance of the date on which they are to be broadcast, they are recorded on either 35-mm or 16-mm film, similar to film used for motion pictures, with a sound track provided. When programs are produced on film, a different production technique is used than that used for live television or cable programs. The various portions of the program are shot in short "takes," not necessarily in the order they will be presented when aired. This involves a technique borrowed from motion pictures. Motion picture cameras, rather than television cameras, are used for outdoor scenes, where the producer is not tied down to using the form used with live productions. Furthermore, programs on film last indefinitely. After a successful network run, they can be syndicated and used over again. From the original negative, hundreds of prints of the complete filmed program may be made, if desired. However, one form of filmed program to which this does not apply is the program on the old kinescope form. An old kinescope film is a motion picture film, either 16- or 35-mm, shot from the face of the kinescope picture tube in a television receiving set. The original program presumably was a live production. A few kinescope copies were made for use by network affiliates unable to carry a network program at its regular network time. Sometimes this technique is used to permit a delay in showing either a network or local program. Use of these old "kines" has been largely replaced by the use of videotapes.

Transcribed Radio Programs

Transcription is the radio equivalent to the use of film. A transcribed radio program is one recorded on discs, twelve inches in diameter, played on turntables rotating at 33 1/3 RPM. Before the days of television and cable, a substantial number of radio programs were available to stations in transcribed form. There is still some use of syndicated transcriptions, and many programs provided for radio by nonprofit organizations are supplied in transcribed form. As in the case of film, hundreds of "pressings" may be made from the original master, corresponding to the original film negative. However, the transcription recording process is decidedly different from that used for film. The entire program is produced in the same manner as a live program and recorded on the original master disc. Obviously, editing is difficult to use in the case of transcription. Incidentally, "Tx" is the common abbreviation for the term.

Radio Taped Programs

Many broadcasters desire to keep a file copy of a radio program or to provide limited numbers of copies of the program for use on other stations, so they record the program on tape. Tape was introduced to

American radio after the close of World War II. The taping process was initially introduced and developed in Nazi Germany during World War II. One device involving use of the radio tape was the "beeper" recording of telephone conversations, and it was widely used by radio stations in the 1950s and 1960s. Entire programs now are rarely prepared in beeper form. However, short interviews secured by beeper have been used on news programs. The term gets its name from the beep at intervals of some fifteen or twenty seconds, allowing participants in the telephone conversation to know it is being recorded. The beeps are naturally heard on tape, when the material is played back.

Video-taped Television Programs

The videotaping process was perfected and introduced to the television industry in 1957. Both video and audio portions of a television or cable program are recorded on magnetized tape approximately three inches in width. The practice of videotaping a television or cable program involves off-the-line taping of a television program being produced live, to permit broadcast at a later hour or a repeat broadcast. Videotape has one major advantage over film for television or cable use. It can be played back within a matter of minutes, while film requires time for development and printing. There is some use of videotape programs for syndication in cases where only a few stations wish to make use of the recorded program.

PROGRAM SOURCES

On the basis of who produces the program, there are several types of production sources for programs. Six of these follow:

Network Production

The program is produced by the network; consequently, all producers, writers, entertainers, and personnel are employees of that network. Programs may be live or use any of the forms of recording mentioned.

Package Agency Production

In this situation, the program is produced by an independent agency specializing in production of programs for broadcast. Programs may be live, filmed, taped, or transcribed.

Non-profit Organization Productions

In this case, the nonprofit organization—a college or university, religious organization, fund-raising organization, United Way agency, or holder of political office—is responsible for producing the program. It employs a producer who is an expert in the field, allowing him to handle all details.

Motion Picture Producing Company Production

Obviously, theatrical feature films and motion picture short subjects, cartoons, and comedy specials are produced by motion picture production companies. If a motion picture studio engages in production of filmed programs for television or cable use, rather than for original use in movie theaters, that studio is classed as a package agency, as far as filmed-for-television programs are concerned.

Advertising Agency Productions

A few network and cable television programs and local radio or television programs are produced by advertising agencies. The agency follows the procedures mentioned for other producers.

Local Station Production

Here, the entire process of planning and producing the program is handled by employees of an individual single station. The programs are local station productions, whether on radio, television, or cable. In a few instances, a program produced by one outlet may be taped and made available for use by others on a syndicated or no-cost basis.

BROADCAST ORIGINATION SOURCES

There are four types of origination. This does not, however, refer to who first thought up the program, but to the source from where it reaches the outlet that broadcasts it. These include the following.

Network Origination

A network origination is a program delivered to stations by means of network lines. Telephone lines are used in radio, while coaxial cable or microwave relay is used for television, and a "dish" receiver from a communication satellite is used in the case of cable. The network-originated program may be network, package agency, or local-station produced. However, it comes to several outlets, which make up the network that broadcasts it by means of AT&T long-lines. There is an exception; if a station is unable to broadcast some particular program at the hour the network feeds it to the stations, a kinescoped or taped version of the program may be shipped to the station by mail or express for broadcast at a later date. It might be two weeks later, other than the day other stations air the program. In spite of delivery, it is still classed as network originated.

Syndicated Programs

A syndicated program is one prepared on film, tape, or in transcribed form and not carried over network facilities. It is provided to the outlet on a rental basis by a syndication or sales distribution firm. The film, tape, or transcription is sent to the station by mail or express. No network lines are involved. As a rule, the station or cable facility broadcasting the syndicated program will return it to the syndication company. In case of theatrical feature film, the same film may be shown two or three times before being returned, depending on provisions of the rental contract.

Programs Provided by a Nonprofit Organization in Filmed, Taped, or Transcribed Form

This is the same as a syndicated program, except no rental fee is paid by local stations for the privilege of broadcasting the program. There may be a small fee charged, but if so, it is one high enough to pay only the costs of packaging and shipping. This type of origination is often classed as a form of syndication.

Local Origination

Here, the program does not come from any outside source. It is produced by the station's own staff, or by a package agency, advertising agency, or employees of a nonprofit organization. It is produced in the local station's own studios, while no form of shipment from any other source is involved.

CLASSIFICATION OF BROADCAST TIME

In classifying station or network broadcast time, various time periods when programs are broadcast fall into two major classifications. On the basis of desirability, some hours of the day are more desirable than others. From the standpoint of the audience available and who tunes in, a period of 7:00 P.M or 8:00 P.M. is much more desirable than 4:00 in the morning. Station practices in classifying time vary widely. However, this illustration is typical and applies to all stations uniformly.

Class "A" Time

This includes the most desirable time periods during the broadcast day. Periods when the station, cable facility, or network makes its highest charge for program time in carrying spot announcements are: for television and cable facilities, that period between 7:00 and 10:00 P.M. (prime time); for radio, which has little evening listening, between 7:00 A.M. and 9:00 A.M. and 4:00 P.M. and 7:00 P.M. in the afternoon and evening (these slots are called "drive times").

Class "B" Time

This is the second period of time in desirability from a sponsor's point of view. On television and cable, Class "B" includes Sunday afternoons from 2:00 to 7:00 P.M., and weekday or Saturdays between 6:00 and 7:30 P.M. Oftentimes, the half-hour period each night between 11:00 and 11:30 is also included. On radio, Class "B" time includes all daytime hours from 6:00 A.M. to 8:00 P.M. that has not been designated as Class "A" time. The Class "B" rate is usually two-thirds the rate charged for Class "A" time.

Class "C" Time

On television and cable, any time during daytime hours, except Sunday afternoons, between 7:00 A.M. and 6:00 P.M. is considered Class "C" time. On radio, Class "C" time includes all broadcasting hours not included as Class "A" or "B" time. Rates charged run one-half the Class "A" rate.

Class "D" Time

On television and cable, this includes all remaining hours during which the station or cable facility is on air. On radio, there is no Class "D" time. Time charges for Class "D" time are about 50 percent of that charged for Class "A" time.

REVENUE-AND NON–REVENUE-PRODUCING TIME

On the basis of revenue-producing or non–revenue-producing status, stations consider broadcast time most important. In addition, the

Federal Communications Commission, despite the great measure of deregulation taking place in programming, expects stations and cable facilities to devote part of their air time to programs where no revenue is received. The difference in types of station time do not refer to short, station-identification periods between programs, but only to time occupied by programs themselves.

Commercial Time

This means any time, other than a station break, on a station, network, or cable facility that is sold to an advertiser for a commercial program, or where commercial announcements for one or more sponsors are inserted. This is the revenue-producing program time.

Sustaining Time

This includes any program time which produces no revenue for the broadcast or cable entity. It is time used for a program that includes no commercial messages, while the station receives no payment.

The placing of a paid announcement in the station-break period immediately before or after the longer period occupied by a program does not make an otherwise sustaining program commercial. If there are no commercial announcements within the program itself, the program is sustaining, regardless of what materials precede or follow it.

CLASSIFICATION ACCORDING TO SPONSORSHIP

The classification of commercial programs on the basis of the type of sponsorship must also be considered. Five types of commercial programs, not including those that are sustaining, come from the number and type of advertisers whose messages are carried, and the manner those announcements are handled.

Regular One-Sponsor Sponsorship

Here, the entire commercial one-time program, or entire series, is sponsored by a single advertiser, who buys the time necessary for presentation of a program and provides and pays all costs of production of the program itself. This is very rare today on network programs, but it does occur on local stations.

Alternating Sponsorship

This situation exists when two different advertisers take turns sponsoring the programs in a series. One week, sponsor A pays the costs of time and production for the program. Consequently, the program carries his advertising messages. The following week, costs are paid by sponsor B, and the program carries his messages. In addition to advertising messages for the week's prime sponsor, courtesy announcements for the nonweek sponsor are inserted in each program. This type of sponsorship exists only on network or cable outlets. It is a direct result of very high costs of sponsoring weekly programs on television networks and cable facilities.

Co-op (Cooperative) Sponsorship

This is the type of sponsorship used when a network or cable system provides a program and feeds it over facilities to a number of affiliated outlets with public service announcements in spots where commercial announcements are inserted. Each affiliate receiving the program has the privilege of selling it to a local sponsor and inserting the local commercial announcements for that local sponsor, instead of broadcasting or cablecasting the network-provided public service announcements (PSAs). If the program is sold to a local sponsor, the station is expected to make some payment to the network toward costs of providing the program. If it is not sold, the station may carry the program with PSAs instead of commercial announcements. If it is a sustaining program, no fee is paid to the network. Co-ops are heavily used on radio and television, especially for network news broadcasts. The term *cooperative* is also used to refer to the situation where a national advertiser and local distributor share costs of local sponsorship of a program or of local use of spot announcements.

Segment Sponsorship

This type of sponsorship is used when a program is divided into fifteen-minute segments, and each segment is sold to a different sponsor for a flat sum that covers cost of time, and a prorated share of the cost of producing the program. The device was first used around 1941, when the hour-long, five-day-a-week network program "Breakfast Club" was divided into segments for sponsorship. Most five-day-a-week programs on networks are sponsored on a segment basis. Most network or local broadcasts of longer-than-average sports events are segmented. Frequently, very expensive television or cable specials will be handled on a segment-sponsorship basis, as well.

Participating Sponsorship

Here the network, cable facility, or station produces and pays costs of the program. They then sell the right to insert commercial messages or spot announcements to a number of different sponsors. The charge made for each spot announcement is between sixty and seventy-five percent of the amount charged for time with a fifteen-minute program. Participating sponsorship is the form used for all disc-jockey or platter programs on radio. It is used with many thirty-minute or longer local radio shows, as well as television, or cable farm information, or women's interest programs. It is used by radio networks in programs such as Paul Harvey's News Program. It is used also by television and cable networks for low-budget early morning, or late-night variety programs, such as "Good Morning, America," "Today," "Tonite," or "David Letterman's Late Night Variety Show." Sometimes it is used on very high-cost specials. It is frequently used for local or regional network broadcasts of baseball, football, or basketball games. Participating sponsorship is the rule with local television showings of theatrical feature films.

CLASSIFICATION OF ANNOUNCEMENTS

Classification of announcements should be discussed, since the real purpose of many broadcast programs is accomplished by use of separate announcements in or between programs. The major types of announcements include the following.

Program Commercial Spot Announcements

These announcements are distinct from the entertainment portion of the program. They are designed to sell the sponsors' products, establish trade names, as well as establish a public-relations form of an institutional advertising campaign. These are the announcements used in sponsored programs.

Institutional Announcements

This is institutional advertising, which is a form of public relations advertising for major corporate entities. These are non-product-selling announcements selling the listener or viewer on the company's good name.

Visual-only Commercials on Television or Cable

In some programs, the name or logo of the company or advertised product is placed on a backdrop, or the front of a desk used by a quiz program master of ceremonies or a newscaster. It also can be superimposed very briefly over lulls in action of sports broadcasts. These are commercials, nonetheless, but are not presented by audio or "voice-over."

Noncommercial Spot Announcements

All stations must carry announcements of a public service nature for charitable organizations, schools, religious organizations, and United Way agencies; these are noncommercial spot announcements, as required by the FCC.

MISCELLANEOUS DEFINITIONS

Delayed Broadcast

This is a program recorded on tape and broadcast by an individual network affiliate at a time later than the hour the program originally was presented and fed over network facilities to other affiliates. The term can be applied to broadcasts or cablecasts by a specific station or entity. "DBs" occur usually in one of two circumstances: one, during daylight savings time when networks "feed" a program at a usual time to stations. However, for the benefit of stations in nondaylight-saving areas and cities, the taping of a program will be made, usually in Chicago. The taped version is fed to standard time cities an hour later than the original feed. The second circumstance concerns only one station, which cannot broadcast a program at a regular timeslot. That station may tape or delay the program until a later time.

Station Break

This refers to the period of time between any two network or cable programs, and between any two programs of any type carried by a station where time is provided for the required station identification. Though the ID may require only a few seconds, the station break period is always thirty seconds in length. In the case of some network programs, an even longer break is made between programs. The objective is to allow station time to insert a greater number of between-program spot announcements. The term *chain-break* is the same as station-break, with respect to network and cable operators.

Augmented Audience Reaction, Live Studio Audience

In the closings of some programs, an announcement is made that "audience reaction has been mechanically reproduced" (augmented) or "This program had been performed before a live studio audience," the former meaning that taped or transcribed applause or laughter or applause has been used, while the latter means that response has come from a live, in-studio audience viewing the presentation. Both are used to increase total volume of laughter and applause heard by broadcast audiences.

Chapter 6

Program Form Considerations

As programs developed over the years, certain considerations were established with respect to forms used for programs. The most important of these are discussed here.

USE OF PROGRAM TITLE

First, unlike the situation in early days of radio, every program put on the air today has a title. It is a label of identification. Exceptions are the broadcasts by the president of the United States. The label serves the function of identifying the program or program series. It also serves as a form of advertising or promoting the program, and attracts an audience. To be highly effective, the title of a program meets four basic requirements.

It should indicate the nature of the program. From seeing the title in a newspaper listing, or having it announced on radio, television, or cable, the individual should have a good idea of what to expect. Such titles as "Gunsmoke," "The Twilight Zone," "Dynasty," and "Bugs Bunny Cartoon Show," are excellent from this point of view.

Titles such as "The Barbara Mandrell Show" or the "Tonite Show" have value only if the viewing person is acquainted with the type of program where Johnny Carson or Barbara Mandrell are associated. In the case of these two entertainers, it is probable that practically all of the viewing public recognizes the names, but the same would not be true in the case of "The Gary Sievers Show" or "The Richard Shortridge Hour." Such program titles as "Alcoa Presents" or "Hallmark Hall of Fame" give the viewer no idea what type of program is referred to by the title.

It should be distinctive and easily remembered. It does no good for a program series to have a title if viewers can't remember it or recognize it when they see it in a newspaper listing. Some titles stick in the viewer's mind—titles such as "Three's Company," "Movin' On Up," "Magnum, P.I.," or "The Dukes of Hazzard." A title suggesting action or having dramatic value is good. A title that is alliterative, such as "News Nightline," is good, as is one in which words rhyme.

It should arouse interest or curiosity and have the quality of making viewers unfamiliar with the program want to tune in. From this standpoint, "Romper Room School" is a good title, even though it doesn't meet the first requirement. "Morning Melodies" tells us about the radio program but lacks action. From interest-arousing standpoints, consider the effectiveness of such titles as "It Takes Two," or "All Creatures Great and Small," or the old Groucho Marx show, "You Bet Your Life," as compared with "The Dick Clark Show," "Friday Nite Live," or "The Joanie-Chachi Show."

Finally, the title must be short enough to go into a newspaper listing. As a rule, newspaper listings of programs provide room for not more than fourteen to sixteen letters or spaces in Century-9-Medium, or PR-10-B type-size. If the title is longer, the newspaper entertainment editor will simply cut words out of the title or substitute his or her own. For example, "Information Please," a very successful program and the first panel quiz type, was nearly always listed as "Information," and that drove the potential audience away. "Too Close for Comfort" is too long and is listed simply as "Too Close"; that gives little information about the nature of the program. Sometimes, a long title is omitted entirely, and in its place, the words *Music* or *Drama* are inserted. Aside from newspaper listings, a short snappy title is more effective. If the title includes the name of a sponsor, or sponsored product, newspapers regard the use of the name in the title as "advertising." The sponsor's, or product's, name is completely dropped. So, G. E. Theatre becomes "Theatre," or "Alcoa Presents" is printed as "Presents" or "Drama."

EXACT TIMING

Programs are handled to run exactly in minutes and seconds. They never vary! The practice of exact timing developed from radio networks and stations. Any given station at that early period could alternate local or network programs. If the local program ran too long, listeners would miss the beginning of a network program, and vice-versa. On the other hand, if the program was too short, there was a period of "dead-air" following that program and preceding the start of the next one. That resulted in listeners turning to other station, which was even more objectionable to broadcasters! So every program on a network today, including filmed, taped, or transcribed programs, as well as live productions, are timed to run exactly the length allowed.

The length will always run thirty seconds less than the announced length. For example, a sixty-minute program is timed to run exactly fifty-nine minutes and thirty seconds, or a thirty-minute program is timed to run exactly twenty-nine minutes and thirty seconds. The thirty-seconds-between-program rule applies with respect to all programs —filmed, taped, recorded, or syndicated. In some cases, program length is more than thirty seconds, to permit insertion of additional station-break commercial announcements. However, if more than a thirty-second cut is made in program length, the program is still timed to run for exactly the minute and second scheduled.

THREE-PART STRUCTURE

This practice developed in radio in the 1920s for building a program consisting of three recognizable parts: a program opening, middle portion or "program proper," and a program closing. The opening and closing are very short. Combined, they use only a minute or two of total running time for the program. The use of program openings and closings originated with radio. However, they are not used in stage dramatic productions or the vaudeville variety that also were developed here in this country. Accordingly, a program opening and closing were never used in motion pictures, aside from an opening title, until years after their use had become standard on radio.

For the functions of the program opening, there are two basic considerations. It identifies the program coming and attempts to capture the attention of the audience. It also serves a third purpose: It sets the mood of the program to follow, as in the case of the opening of the "Dukes of Hazzard," or on the old "Lights Out," "Danger," or "Inner Sanctum" shows.

To identify it, the program opening tells the viewer the title of the program, identifies the featured personality, and gives names of any featured guests to appear on that program. Occasionally, supporting personalities appearing on a regular basis are also identified. Anthologies or other types of dramatic series may give only the title of the particular dramatizations to be presented, in addition to the title for the series.

To create interest, the opening depends on the identification provided. The enthusiastic style of the announcer giving the name of the featured personality is a good example. Many adult westerns, and now prime time comedy or adventure series, open with a short humorous or suspenseful scene from the program to follow before the program title is given. Some quiz programs open with a quiz question and answer given by a contestant. These "teasers" are all part of the program opening.

A majority of programs make use of "signature" material in the program opening. A variety of signatures can be used, including musical interludes, spoken lines, sound effects, or a visual. To illustrate, the opening of the "Tonite Show" always includes an announcer giving location, guests, orchestra direction, and an orchestration that is a very short strain from "Johnny's Theme." The old "Henry Aldrich Program" on radio opened with a woman calling "Henry! Henry Aldrich!" followed by a boy's voice asnwering, "Coming, Mother!" "Gangbusters" started the opening with a series of sound effects including marching feet, a police siren, machine-gun fire, and sounds suggestive of prisons. This opening was transferred to television when the series appeared there, with each particular sound effect matched to corresponding visual effects in short "takes."

Every program, including local productions, begins with a standard film clip, including visual and audio materials, without change, as the standard opening for every show in the series. In some cases, actual visual material changes with each broadcast, but it essentially works to the same climax. For example, in the opening of the popular "Hee-Haw" program, a different filmed cartoon involving country or rural settings are used. Even if specific materials are different, there is enough commonality to each opening that the material is still classified as a signature.

In analyzing the functions and content of the program closing, this vital portion of the program serves two functions. It fills out the program and gives a feeling the program is formally completed. It also provides final identification for the program and those who contributed to its production. Bringing the program to an end does not refer to a final portion of the program proper. It might include a production number for a musical program, or the last big scene in a dramatic presentation. The closing of the program begins after the program proper has been concluded.

The identification function of the closing includes a restatement of the title, so that those who tuned in late may know the identity of the program they have been listening to or watching. In many programs, the identification section includes an elaborate listing of credits. On radio programs, before the rise of television, credits were given orally, by the announcer, to important members of the cast or entertainers who appeared. This included a final credit to the featured personality. The entertainers being credited came about as a result of union rules. There are no printed programs in broadcasting or cable; nevertheless, actors want recognition. In network television, guests, cast members, directors, producers, camera directors, set designers, or writers, as well as those who handle lighting, makeup, and costuming receive credit.

Frequently, a closing signature is used. If a musical signature is used in the opening, the same music is used "over," or "under," spoken or visual materials. Two elements included in closings are a "pad" and a "tease." A pad is simply material presented rapidly or slowly, or a technique whereby materials can be omitted, such as cast credits on a radio program, or a musical theme repeated several times. The pad is material that can be contracted or expanded to insure the program ends exactly on time. Pad materials must be included in every live production on radio, television, or cable. However, if a program is filmed, no pad is needed. The program is cut to exact timing by cutting out film footage. Taped or transcribed programs usually make use of a pad. Such programs are normally produced in the same manner as live shows.

On television, a frequently used type of pad is listing all credits. In live programs, they are mounted on a drum-shaped, electronic, "crawl" machine, which is turned rapidly or slowly, as desired, to develop exact timing.

Secondly, an additional element in a program closing is a "hook," or tease, for either the next program to be broadcast in the same series, telling who the guests will be, or the progam to follow, on the same station.

Chapter 7

Program Idea, Featured Personality, Program Formats

This chapter deals with three aspects of radio, television, and cable programs that are vital for its success: the program idea; featured personality around whom the program is built; and the format used on the program. The first two require a considerable amount of discussion, while the third can be discussed briefly.

THE PROGRAM IDEA

Every program or series is built around an idea. The idea is systematically developed until it becomes the basic framework of the program.

Developing the Basic Idea

A clear picture for the place of an idea in a program should be given emphasis by tracing development of a program idea from inception until rounded into the foundation of a program or series. For illlustration, let us consider development of the idea for a series rather than for a one-time special program.

Step A

To begin with, someone comes up with an idea for a program series. It may be a very simple idea. For example: "Let's put together a local quiz program to fill that slot at nine-thirty in the morning:" or, "We want a space adventure, or something off the beaten track"; or from a time salesman, "This advertiser wants to sponsor a program of recorded music, easy-listening, but not top 40." That is the initial beginning of the basic idea. The real work of developing a program will come later.

Step B

The next step is to elaborate on that basic idea. Future steps in the development process may be taken by a single individual or by a variety of individuals in a series of discussions and arguments over the form the program should take.

At any rate, questions must be answered, such as if a quiz program, what kind of prizes should be given? If a live musical program, built around a specific vocalist, what kind of format should be used? What sort of instrument support should be given the singer? What use is made of other vocal groups, to provide relief from a long series of solo numbers?

From this point on, we can take an actual program, such as the long-running "Gunsmoke" series. The persons who planned that program started out simply with the idea of producing a western dramatic series. What kind of western would be attractive to viewers and still be different? The possibility would be a program where human-interest values would be emphasized to a greater degree than in existing westerns of the type, such as "The Lone Ranger" or the old "Roy Rogers Show." Therefore, the real program would be aimed at adults rather than at young children.

To that end, characters must be real, true to life, and believable. To add to the feeling of reality desired, all episodes would be presented in the locale of an actual town, Dodge City, Kansas, whose early history in 1870 through 1880 would provide the opportunity to introduce action in the series. While the locale would be historical, actual plots would be fictional, so the series would not be bound by limitations adhering to historical happenings. For such a series, a thirty-minute or sixty-minute running time would be most appropriate. The difference is provided partially in the setting. By providing a sense of realism and reality, the program gives emphasis to human interest values.

Step C

The third step is selection of a featured personality, around whom the program would revolve. In a quiz program, this would be the master of cermonies; in a variety program, the host; and in a platter program, the featured disc jockey. No matter what the type of program, the person who is the featured personality will be highly important. In dramatic feature programs, every episode is built around activities of a central character, and he is the core throughout the entire series.

Using the "Gunsmoke" illustration, the first important question is, "What kind of central character should be featured?" In line with earlier decisions to emphasize human-interest values and problems, the central character has to be a rather ordinary person. He cannot be a dashing, romantic figure nor a great lover or hero. He has to be an everyday person, capable of making mistakes. To justify his interference in every episode and affairs of the other people, the "Gunsmoke" central character was cast in the role of a peace officer, a United States marshal, headquartered in Dodge City.

Step D

Next, what regular support could be given this central figure in each program? In the case of a quiz program, the master of ceremonies would need one assistant on stage, in addition to contestants. In a live television, or cable, network musical, the central personality would be supported by an orchestra, a vocal chorus, a group of dancers appearing in every broadcast, as well as stars for each program. In dramatic presentations, the central character has one or more regulars who appear each week.

In the case of "Gunsmoke" several such regular subordinate characters were necessary. If the program were to stress human-interest values and make the hero a real person, it had to include someone the hero could talk to about his problems. Again, with human-interest angles in mind, the central character in "Gunsmoke," Marshal Matt Dillon, was given three regular supporting characters: Chester (originally—later Festus Hagan), the not-too-bright character, assistant to the Marshal; Doc, a grumpy, complaining frontier doctor; and Kitty, a somewhat idealized dance-hall hostess. Note the fact that each of these supporting characters, or regulars, contrast strongly with the character given the marshal and with other regular supporting characters in the program.

Note also, that each of the three supporting characters offer excellent possibilities in direction of adding human-interest values. This was done in keeping the decision to make "Gunsmoke" a program with heavy human-interest appeals.

Step E

The final step is working out details. The format or time pattern and transitional music to be used in each broadcast in the series must also be considered. Selection of an attractive and appropriate title, development of an effective opening, as well as a standard closing should be considered. In the case of a quiz program, the type of audience participants to be used and how they would be selected are considered. In part, this final step is a matter of working out necessary details with respect to the program. It is a polishing process, by means of accentuating different qualities provided by the program. But in every situation, details are worked out with the basic purpose of making the program more effective and attractive to listeners or viewers.

These five steps are those followed in developing the idea of any program. After the program's concept has been given final form, it is turned over to a producer or producer-director, who finds the right person to serve as featured personality, employs a writer or writers, and proceeds to put the program on the air. In discussing the program idea, we are not referring to the original part of an idea whence the whole thing started, but to the fully developed one, as it would be after completion of all five steps in the program development.

THE FEATURED PERSONALITY

Most, although not all, radio, television, and cable programming is built around a featured personality. For purposes of this book, we do not regard the leading man of a single broadcast in an anthology dramatic series as a featured personality. We reserve the term as referring to the central figure in each broadcast in the series. If the program is a "one-time special," the term *featured personality* can be applied to the individual who is featured in that one broadcast. The limitation imposed applies only in relation to a program series, rather than to a one-time program.

Who is the featured personality?

He is the featured newscaster on a news broadcast; the disc jockey of a platter music program; the featured comedian on a comedy-variety series; the master of ceremonies of a quiz, audience-participation, or human-interest program; the host/master of ceremonies of a vaudeville-variety type program; the central hero or heroine of a dramatic series built around the same person each week; or he could be the host for a dramatic anthology series where a different guest star is the "lead" each week.

In some cases, as in news, platter, and vaudevile-variety programs, the featured personality appears as himself, under his own name. In other cases, he plays a role, as James Arness played Marshal Matt Dillon in the "Gunsmoke" television series.

Role-playing by Featured Personality

In reality, role playing is not limited to the featured personality appearing as a character in a dramatic program. A good deal of role playing is present even when the featured personality appears under his own name in a nondramatic program. Certainly, entertainers on the air look their best in front of a camera or microphone. Sometimes, they cast themselves in a role considerably different from the one that seems natural to them, as long as they are on the air.

Requirements of an Effective Featured Personality

Whether role playing or not, the successful featured personality must have certain definite and recognizable qualities, in character, when appearing before the microphone or on camera. First, he must be vital and alive. A "tired" personality may be presented as a character in a comedy drama program, but that is unusual. In every other situation, the featured personality must be alive, vital, and vigorous. They must give that impression to listeners and viewers. If he is unwilling to work to entertain, the audience won't be willing to tune in that program.

Second, he must be vocally and physically attractive. This doesn't mean that he must be handsome, or have a nice manner of speaking. It does mean that the voice should be pleasant and warm and his diction good, to mark him as a moderately well-educated person who is careful of speaking mannerisms. Physically, if he appears on television or cable, he should be of good appearance—neat, not careless in dress or in the way he conducts himself before camera.

Third, he must be intelligent. Listeners and viewers will not respect a person who makes it evident he lacks ordinary common sense or is below normal in intelligence or education. Obviously, it is possible to go too far in the other direction, but that danger is handled by other requirements.

Fourth, he must be an identification symbol for the audience. Audiences do not respond to snobbish featured personalities. We want personalities to be like people who live next door to us. The intellectual who parades his intelligence, the successful comedian who lets his success go to his head, and the Ivy Leaguer who simply can't forget his Harvard or Princeton background and goes to considerable length to impress it upon other people will not register with the audiences they must face.

Finally, he must have the ability to make listeners and viewers respect him and like him as a person. He must be pleasant, not irritable, when things go wrong. He must be able to take things seriously when the sitution warrants such action. He must give the impression of being honest, straightforward, and sincere in everything he does, unless such honesty implies a lack of tact in dealing with other people. He should have a genuine quality of humility. Certainly he's a success, or he wouldn't have a show of his own. He should know that his success, or ability as an entertainer, doesn't put him on a level above that of other people.

He cannot talk down to listeners and viewers. A "know-it-all" attitude repels audiences. He should be the type who really cares about and likes people. This implies the identification term previously discussed. Through identification, this character should go to considerable lengths to avoid hurting people or their feelings.[1] But on the other hand, he should be interested in people and their problems, while trying to help them. He does not have to be brilliant or perfect, but he must be human enough to make mistakes and admit it when he does. The closer he can get to identification, the more effective he will be as a featured personality. All five of these requirements are also important for candidates for public office who appear on political radio, television, or cable casts.

There are some exceptions to the list of requirements outlined. We've had our Walter Winchells, Jack Paars, Jackie Gleasons, and Johnny Carsons who have been highly successful. But failure to meet all five of the requirements disqualifies the entertainer as an effective personality.

THE "DIFFERENT QUALITY" OF A PROGRAM

Programs are based upon ideas in other programs. It is very unusual when an entirely new program form is developed. Entirely too many programs are quite evidently copies of programs on other networks or stations, complete in every detail to attract listeners and viewers.[2] Really effective programs, however, aren't copies of other programs. Every effective program on radio, television, and cable has some quality or qualities that set it apart, as distinctive and different from others of the same type. They can be referred to as that "different quality" in a program.

Sources of difference in a program can be achieved in a variety of ways. It may lie in selection of materials used in the program. The old "Playhouse 90" program and "Hallmark Hall of Fame" made use of original ninety-minute or two-hour scripts, while the old "Lux Radio Theater" program of prewar radio networks days used only dramatizations based on adaptation of feature motion pictures. Other sources of difference lie in unusual structures of programs. As in the old "First Nighter" program on radio, the play itself was surrounded by a framework suggesting the first-night run at a Broadway offering in New York.

Other sources may be found in the unusual mood, or style, of presentation, as the case with Alfred Hitchcock's dramatic series, featuring "chiller" suspense, or low-key acting characterized by radio and television versions of "Dragnet" and "Badge 714," by the late Jack Webb's Mark VII Limited Production Company. Unusual moods and characteristic dialogues set these programs apart from others of those genres and eras. Still, other qualities lie in the personality of the featured individual, as illustrated in the old "Jack Benny Program," or any other name comedian. The same is true of many national network programs built around featured comedians or masters of ceremonies. It can also be found in the unusual setting, with examples ranging from radio's old "Duffy's Tavern" series, where all action took place in a bar in New

York, to "The Alaskans," where the setting was Alaska, or today's "The Love Boat," which captures a distinct setting—episodes take place on a pleasure cruise liner.

It can be found in sheer elaborate and lavish production, as would be the case of many network color spectaculars, and of money giveaway programs such as Monty Hall's "Let's Make a Deal." It can also be found in the use of special feature spots or gimmicks, as in the old "Pot of Gold" program, where a five-minute spot was inserted in a thirty-minute program of popular orchestra music. In that spot, a telephone call was made at a telephone number chosen at random, with the person answering receiving $1,000.

This illustrates a few, but not all, of the methods used to make a program different. In many cases, several different elements combine to provide difference. It is not merely the distinctive style of the featured personality, but other elements as well. In analyzing a program, look for elements that make a program different. If there are none, it is likely to be a low-audience preference.

The format of the program is most important. Most programs presented on a series basis follow much the same pattern in every broadcast. Materials may be different, but structure remains very much the same. Practically every program in the late-night evening network television production "The Tonite Show," featuring Johnny Carson, uses this time pattern:

0:00 Audio background with Ed McMahon, with visual artwork still and action opening; straight cut to Johnny Carson.
1:00 Johnny Carson and opening "live" comedy mologue
7:00 First commercials, (three national and two local, with local options)
9:30 Carson and McMahon dialogue (Burbank, California backdrop)
12:00 Sketch (comedy routine with Carson and McMahon)
19:00 Second commercials (two national and three local, with local options)
24:00 Carson introduces first guest: talk-entertainment (patter)
30:00 Third commercials (options, network-local)
35:00 Carson introduces second guest (or group)
35:30 Second guest (or group) does lively production number with NBC orchestra under baton of "Doc" Severinson
40:00 Second guest (or group) with Carson: patter
47:30 Second guest (or group) does lively second production number

with NBC orchestra, "Doc" Severinson
50:00 Fourth commercials (network-local options)
55:30 Carson introduces third guest
58:30 Fifth commercials (network-local options)
59:30 Program closing

The "Tonite Show" does not adhere rigidly without exception to the pattern listed. If the guests are comedians and scheduled to do a monologue, the time pattern is shifted. There is a considerable amount of elasticity permitted in time devoted to various segments of the program. However, the pattern given is a usual arrangement. It provides a point of departure for time patterns used on all of the "Tonite Show" programs.

A regular time pattern, listing in order the various segments making up any typical program in a series and the approximate starting time of each in minutes and seconds from a start of that program, is called the "format." Using a well-standardized format has two advantages. First it gives the producer, or writers of the program, a tested formula to be followed, along with various limitations and variations permitted, for each broadcast in the series. Secondly, it allows listeners and viewers to know what is coming when they tune in. It will also give something familiar for the audiences to recognize. When a listener or viewer recognizes familiar things, his sense of participation or involvement is increased. Even in the case of a one-time program, a format is generally worked out before writers attempt to prepare an actual script.

Describing the idea of a program is needed to understand formats. If you are asked to describe the idea of the program, remember the description should be of the program series, rather than an individual program. You will need to know what that program is like every time it is broadcast. Consequently, the following information is important:

a) length; frequency of broadcast; evening or daytime; origination; and general type of program
b) featured personality; name and/or name of character portrayed, with a twenty- to forty-word characterization
c) supporting regulars, characters or individual entertainers, choral groups, dance groups, if any are used, with brief characterization of each
d) if name guests are used, how many on each program, and what type

or types; name two or three who are typical for illustration
e) if audience participants are used, how many on each program; what type or types
f) describe how the program goes in a typical broadcast. If program is dramatic, an outline of a typical plot should be used
g) details: type of music used; prizes given if prizes are used; is the production rather elaborate? Explain.

By way of illustration, the following is a description of the program idea for "Gunsmoke."

"Gunsmoke" was originally a thirty-minute, and ended as a sixty-minute, once-a-week evening television show. The now-syndicated adult western program featured James Arness as Matt Dillon, marshal of Dodge City in the early 1870s. Dillon is represented as a large man, pleasant, good natured, somewhat slow, interested in affairs of other people, and invariably helpful—a man who, as a marshal, wore a revolver but rarely used it. In the series, he was supported by other regular characters. Originally, there was Chester, a not-too-bright handicapped assistant to Dillon. Later, in the longer series, Festus was cast into that same type of character portrayal. Doc, a grumpy, kind-hearted doctor, and Kitty, a dance-hall hostess, were two other characters. A typical program might show Dillon assisting a family of homesteaders, one of whom was wrongfully accused of robbery. While Dillon would enforce the law, it would be in a sympathetic manner and in a way that saw to it that justice was allowed to triumph without gun-play, in most cases. Different qualities about the program are: the personality of Dillon, decidedly unlike that of most Western law-officer heroes; excellent character portrayal of secondary characters; and unusually strong emphasis on human interest values in each broadcast.

A similar description can be provided for any broadcast program series or for any single one-time program that is broadcast, as well.

Notes

1. Professor L. H. Mouat, in "An Approach to Rhetorical Criticism" *The Rhetorical Idiom*, ed. Professor Donald C. Bryant (New York: Ronald Press Co.), 1958, p. 167 points out that "The term identification includes the art of employing topics in such a way that one's proposals are identified with the beliefs and desires of the audience and

counterproposals with their aversions, but it is also more extensive and more intensive than this. . . Identification is a process of becoming substantially one with the audience. It is an attempt to proclaim a unity among men at odds with one another . . . but content is paramount. The audience is given the material it wants and needs. In other respects regarding the term "identification," Professor Kenneth Burke, in *A Rhetoric of Motives, A Grammar of Motives*, and *A Philosophy of Literary Form* points out that persuasion is use of language symbols that persuade through the concept of identification. In other words, when one identifies with someone or something, he is becoming consubstantial with that person or thing. This is the significance of the term throughout this book.

2. A somewhat close similarity to this program series, featured a "spin-off" of the successful "All in the Family" program series, entitled "Archie Bunker's Place," where the action takes place in a bar or tavern that Archie Bunker, played by Carroll O'Connor, has purchased. There is a remarkable series of moving from the "All in the Family" series, to the later series of "Archie Bunker's Place," with interconnecting plausibility.

Chapter 8

The Units of a Broadcast or Cable Program

It is a recognized psychological fact that no person is able to give complete and undivided attention to any one thing, idea, or stimulus for more than a very few seconds. Another fact in relation to broadcasting and cable is that no listener or viewer can give even close attention to any one idea or element in a program for any appreciable length of time. After twenty or thirty seconds, attention becomes less. One can focus attention on sound from radio or pictures from television and cable only by exerting a degree of effort. It could be small effort at first, but it is still an effort that becomes continually greater as time goes by. Very few listeners or viewers like any program well enough, or are interested in any subject or type of material enough, to be willing to work at giving attention for any period of time.

To deal with the serious problems of holding attention, radio, television, and cable programs must, at very frequent intervals, give listeners and viewers something new to hear or see. When something new is offered, attention is renewed. As a result, radio, television, and cable programs are divided into very short units or segments, with each offering a change from what has gone immediately before to something new. In this book, reference is made to these short segments as "program units," or simply, "units." The term has been used in the industry. The words *spots* and *bits* are used with similar meanings and contexts. But since we need a definitely recognizable term, we use units in this context.

NATURE OF UNITS

A unit is a short section or segment of a broadcast or cable program offering where one type of material is presented, or one idea dominates the action. The material or idea is, in some evident way, different from material or ideas immediately preceding or following the unit. It starts whenever there is a change in material or idea sufficiently evident to the listener or viewer. A number of different types of changes are possible in order for a new unit to be started:

1) making a change in scene or setting before the camera
2) changing type of material presented from talk to music, music to dance, or action to talk
3) starting a new musical number after an earlier number has been completed
4) introducing a new idea into talk, making a definite change in subject matter
5) using a different method of presenting the same sort of material, such as a dialogue or interview, instead of a monologue.
6) bringing a new important character into a dramatic scene or into conversational situations
7) changing moods from smooth-running, easy conversation to excitement
8) changing from a talk scene to one of violent action
9) shifting focus on listener or viewer attention from one scene, person, or object, to another

This is not an exhaustive list, because there are additional ways new units can be started, as well. But the list illustrates that a new unit will result from the introduction of that change. The more complete the change, the more evident the beginning of a new unit.

UNITS IN RELATION TO TELEVISION CAMERA SHOTS

A program unit isn't the same as a separate camera shot. Units relate to materials in the program. If the same essential material remains "on camera" in a television or cable show, changing the angle of the shot, by shifting from one camera to another, will not produce a new unit. What determines the start of a new unit is whether or not there is change in material or subject matter presented, or in the way new material or subject matter is brought to the attention of listeners or viewers.

Some types of camera changes, however, may mark insertion of a new or different unit. Suppose material before the camera is a large scene, including a number of people, as in a "cover shot," and the camera "dollies" up to an individual. This, consequently, is a change in subject, and therefore, a change in unit. Or, suppose in a dramatic program, a situation develops where exact time is significant. A short four- to six-second shot of the face of a clock on the wall, showing exact time, is significant; with that sweep, the second hand swings rapidly toward the "explosion" point. That certainly would be a significant change to justify consideration of that shot as a new unit. There must always be a change in essential subject, where attention of the listener or viewer is focused, before a change in camera shots can be considered as producing a different unit.

TYPICAL UNITS

To explain the concept of a program unit, the following are listed and regarded as separate units.

A Complete Commercial Spot Announcement

This marks a change from type of material it follows. The announcement will be regarded as being only one unit, even though several types of material or visuals are included.

A Musical Number By a Vocalist or Orchestra

The number follows other types of material such as a spoken introduction, if more than a sentence in length, or a different musical number.

A Dance Routine, Occupying Only Part of a Musical Number

The number might be presented by a vocalist, and the camera is on the vocalist. Then a dance team appears and attention focuses on the dance team, even though the vocalist continues to sing. This results in two units for that number: one for the time where attention is focused on the singer, and one focused on the dance team.

A Monologue By a Comedian

A prime example is Johnny Carson's standard opening monologue. This may sometimes be broken down into two or even three units, if there is distinct change in setting, method of presentation, or subject matter. However, this must be a very real change.

A Bit between Two Characters

It must involve only two characters, and subject matter, mood, or amount of action must remain the same. If a third character enters the scene, there is invariably a change in emphasis and usually in subject matter. Entrances of new characters start a new unit.

A Major New Item in a New Broadcast

Any time important enough to be given thirty seconds or more of program time.

A Group of Related News Items, Each One Short

In the case of a group of three or more one-sentence items dealing with say, traffic accidents, the entire group constitutes one unit.

A Scene in a Dramatic Show

This means any segment where there is no change in main characters on stage, in place, time, subject discussed, or account of action. If, while one scene is in progress, a character sings a song, then that is a new unit. The same would be the case if, after a moderately long all-talk segment, the situation becomes one involving actual physical violence.

An Interview with Any One Audience Participant, in a Quiz Show

This would include the entire interview portion, but not the quiz questions. In the old "Groucho Marx Show," two participants were interviewed at a time, Marx directed several questions at one, then several at the other. His interview with each one was a unit.

A Quiz Question and Its Answer

Since both questions and answers involve the same people and deal with the same subject matter, they comprise one unit.

This is enough to demonstrate typical units and their situations. Applause or laughter following a segment of entertainment materials will not be classed as a separate unit, but as part of that unit represented by preceding material. A very short introduction, not more than one sentence, of an entertainer or a musical number is not classed as a unit. It will be combined with the unit following. Bridge music in a dramatic program and entrance music are too short to be considered as units.

In most programs, average length of units are from forty seconds to two minutes. In television and cable, the possibility of varying camera shots makes it possible to have somewhat longer units than normally used in radio. It will be very unusual to have a segment shorter than fifteen or twenty seconds that is considered a separate unit. However, it is possible to have, as in the clock-on-television illustration that was given, an entire unit in that segment. In a long talk, significant changes in subject matter justifies the starting of a new unit. For analysis, similarly, a change in the method of presentation is a new unit.

All too often, units in straight talks are much too long. That is one major reason why few "straight" talks on radio, television, or cable are able to hold the attention of listeners or viewers.

Chapter 9

Requirements of Effective Program Structure

If a program is to be effective, it must be easy for a listener or viewer to listen or watch without effort, on their part. That listener or viewer should be able to follow the idea and give close attention. Ease of listening or viewing is possible only if the program is effectively put together. It must use materials that permit it to be a strong program. It should provide, at the same time, those materials in proper proportions and present those materials in the most effective arrangements.

Effective structure demands satisfaction of seven structural requirements. The seven requirements are:

Program Unity
Effective Opening and Closing
Strong Beginning
Variety in Materials
Unit-to-Unit Change
Good Pace
Effective Use of Climax

PROGRAM UNITY

An effective program must have a high degree of unity. It must deal with one idea, and every element in the program must contribute directly to that idea. The mood of the program must, likewise, be consistent. Any element, meanwhile, that fails to fit the mood of the program must be eliminated. This does not mean that only a single kind of material can be used.

A musical program need not be made up entirely of vocal solos by a single entertainer. But the music used, no matter the type or means presented, must harmonize with the general spirit or mood of the program. To use an extreme example, inserting a "Rolling Stones" number in a program of symphony music, or a prayer in the middle of a comedy monologue, simply will not harmonize.

Several factors contribute to program unity. They include the following.

Mood

Use of a consistent single mood or spirit throughout the program, while eliminating all materials that do not fit the mood, creates unity.

Theme

Use of a single theme is essential. A comedy variety program may be made up of segments relating to a specific type of situation, such as taking a vacation, Christmas shopping, or paying federal and state income taxes. A forum discussion, obviously, will center around one phase of a specific problem. A dramatic program will be built around a specific plot situation, involving a problem and its eventual solution. The single theme idea is not a requirement in every program, but when used, it will increase unity.

Personality

Using a single featured personality is desirable. Broadcast and cable programs are built around one individual, a newscaster, featured comedian, master of ceremonies, disc jockey, or actor. Even anthology-type dramatic programs, using a different showcase of characters and featuring different people in each broadcast, must center attention on one character or "protagonist," in each program.

Centering attention on one person obviously heightens unity. In fact, very few programs really attain a high degree of unity if they fail to focus attention on some single individual. Although, at times, sufficiently strong focusing of attention on an idea may take the place of such centering upon an individual.

Transitions

Effective transitions and lead-ins are highly advantageous for the program. This is a highly important factor in achieving program unity. To hold attention, unit-to-unit change is essential. Each unit should lead into one that follows, or a transitional introduction must be provided by a master of ceremonies, or a television or cable visual device inserted, that allows the listener or viewer to know what is coming. In dramatization, the viewer must be prepared for materials that are coming by the planting of devices in those earlier scenes or units.

One aspect of unity important in a program series is called week-to-week unity. By providing some common element in each broadcast in the series, the viewer knows, when tuning in, the general type of program he can expect. All programs presented as part of a series use essentially the same types of material in each broadcast, while having the same mood, following the same basic format, and centering around the same featured personality as the key entertainer or master of ceremonies.

Otherwise, the program presented for that week may have nothing in common with the broadcast presented under the same title the previous week. Consequently, this program may not even appeal to the same kinds of listeners or viewers. This results in the series not really being a series at all. Rather, it is a collection of unrelated broadcasts giving listeners or viewers much less reason for tuning in than the case of week-to-week unity provided.

EFFECTIVE OPENING AND CLOSING

Every broadcast or cable program has an opening and closing section, apart from the program proper. Effective program structure requires use of an effective opening, including program title, and an effective closing. Requirements are discussed here in detail.

Effective Title

This title should indicate or suggest the nature of the program. It should provoke interest, be "catchy," distinctive, easy to remember, and short enough to fit into newspaper listings, where not more than sixteen letters or spaces will be printed. While use of sponsors' names as part of titles help identify the sponsor with the program, in newspaper listings, sponsors' names are invariably omitted. Therefore, remainder of the title must meet requirements given.

Effective Opening

An opening must identify the program. The title must always be given in the opening, where appropriate; the name of the featured personality and announcement of any featured guests should also be given. It must arouse interest, with attention captured at the very beginning, or the set may be turned to a different station or flipped to another cable channel. Video or audio signatures are usually helpful for both identification and arousing interest. Many television dramatic programs use an opening device: a thirty-second selection of the program, showing strong action and providing attention values. Then, the production follows with an opening proper, where title and other identification materials are given.

Strong Start

In either radio or television, it is vitally important to capture the attention of listeners or viewers at once and hold that attention until the program proper is well under way. Obviously, a part of capturing attention is taken care of by the use of effective interest-arousing program openings. But, this alone does not satisfy a strong start requirement. Two additional factors are most important.

The program proper must get underway without delay. A long opening results in a weak start, and obviously a long commercial spot following the program opening weakens the strength of that opening. If the opening did secure attention, the long commercial allows attention and interest to dissipate. If an opening commercial "spot" announcement is used, it should be held at minimum length, with longer time used for commercial materials later in the program. The faster the program can be brought into the program proper, the better it is for that particular programming.

STRONG ATTENTION-BUILDING FIRST UNIT OF PROGRAM PROPER

A variety program should use a lively production number as the opening unit of the program proper. A musical program should use a musical number with full orchestration, rapid tempo, and plenty of action. A news program should open with a highly interesting news item, while talk-show openings should contain some type of attention-getting device. Frequently, audience-participation and panel programs fall short of meeting this first initial unit requirement. At times, the format calls for an explanation, by the master of ceremonies, of how the program "game" goes. This is an obvious weakness, since any explanation should be held, if at all possible, until something has happened. In a dramatic program, the first scene or unit of the dramatization should be an action unit, excluding lengthy explanations.

VARIETY IN MATERIALS USED

This is an obvious requirement of effective program structure. Any program consisting entirely of one type of material is certain to become tiresome to the listener or viewer. For example, a half-hour program consisting of one vocal solo after another by the same vocalist is far less effective and interesting than a program where vocals are mixed with instrumental and group numbers, and solos are presented by trios, quartets, or other musical groups. Some methods of providing variety in materials should be suggested.

Variety in Kinds of Materials

There are different kinds of materials which can be used in broadcast or cable program. For example, in music alone, there are ballads, novelty, soul, rhythm and blues, rock, folk, jazz, Latin-American, Hawaiian, and easy-listening instrumental segments, to name a few. This is only a beginning. A variety program may offer comedy monologues, dance routines, choral numbers, and song-and-patter teams.

Variety in Methods of Presentation

For example, a specific musical number might be presented as a vocal solo, a trio, quartet, or other small group. The number must have the backing of a large chorus, or if it is an instrumental solo, it should be presented by a small instrumental group. It could be presented by a full orchestra.

Talk materials might be presented as straight talk by a single speaker, dialogue between two speakers, or straight presentations by two speakers alternating with a question-and-answer segment. It might include a single speaker in the form of reading a section of a letter or reading from a book.

Variety in Types of Entertainers Appearing

This is well illustrated by types of characters ordinarily found in a simple dramatic presentation. There will be a male lead and a corresponding female lead. It could have a comedy-type friend of the hero; sometimes a juvenile character or teenage boy or girl is occasionally used. Often a younger child can be utilized, as well as one or two older people who are involved in a well structured dramatic offering. There are never important characters of the same sex and age, unless variety is provided by making the characters quite different in other respects.

Television or Cable Variety in Settings

In most television and cable programs, one method of providing variety is that of using a number of different sets.

Unit-to-unit Contrast and Change

This requirement of good program structure should be considered as one aspect of variety. It will add to the program as a direct result of structure, as opposed to selection of materials that are included in the program.

In short, as much change as possible from the unit that preceded it in the program should be provided in each unit, without violating the requirement of unity. An effective program is not built if the format calls for one vocal solo followed by another, or if one song-and-dance team in a variety program follows another. Each unit should be as different as possible, in content, type of material, and type of presentation, from the unit it follows. In addition, care must be taken that no unit too closely resembles any unit presented earlier in the program.

If the unit in question is the tenth in the program, it should contrast strongly with unit nine. It should also, in some way, be different from units one to eight, inclusive.

For example, in a musical program, the first unit in the program proper should be a lively production number presented by a vocalist supported by a choral group, and backed by a full orchestra. The following unit should remain in the music vein, but use a slower number, possibly a love song, featuring a female vocalist, without elaborate production support. Unit number three in the program proper is still a musical number, in contrast with the second number, a lively type, but not to the degree used in the first musical number.

Here a novelty number is appropriate, featuring an instrumental group. Then the fourth musical number in the series must contrast with the third but still be different from the first and second. A ballad by a duet, possibly a man and woman, would fit this requirement. This should provide contrast, continuing throughout the program.

Good Pace: the Effect of Moving

In every type of entertainment, pace is one requirement always indicated. Pace is not, in itself, a synonym for fast numbers. This refers primarily to an impression of going somewhere and not dragging. In some degree, the effect of good pace comes from avoidance of materials that move slowly. A very slow speaking monologuist cannot be allowed before the microphone for more than a very short time, or a very slow musical number cannot be played for too long a period. In considerable degree, too, the effect of good pace comes from avoiding stretching out the materials. Otherwise, listeners and viewers get the impression the program lacked enough material to fill out the time and padding was necessary. That, naturally, interferes with attention and interest.

The most effective test of whether or not a program has good pace is a check on the length of units making up the program. If units are regularly long, pace then is necessarily slow, and consequently, the program suffers. Even if one or two units are unusually long, listeners and viewers receive a distinct impression the program is dragged out. On the other hand, a program where unit changes come frequently, long before listeners and viewers tire of the material in that unit, one will find gives impressions of moving rapidly. In other words, this is a program that satisfies requirements of good pace.

Effective Climax and Building

The final requirement of good program structure is effective use of climax. Listeners' and viewers' interest not only must be aroused, it must be maintained throughout the program. Interest and attention should reach a peak at the very conclusion of the program proper.

In a musical or variety program, final climax is achieved by use of one or two devices. Either the final unit is a big number or elaborate production, or the unit makes use of strong emotional stimulation. Sometimes a comedian can reach a peak of effectiveness by a last-unit appeal to emotions. More often, however, the device of elaboration provides a climax. And of course, in some programs, the format calls for a combination of elaboration and depth by use of a hymn or other religious piece of music as the final unit in the program proper. With music given production treatment with a device characteristic of the old Tennessee Ernie Ford evening television program series, climax will be achieved. In a dramatic program, the denouement or final solution of the hero's problem will generally be the final unit of program proper. The final unit, in this case, will not be the climax. The climax immediately precedes it at the time when everything looks most impossible for a solution.

Since the climax should come at the end of the program proper, this type of final solution should be kept as short as possible, with a minimum of explanation offered. In quiz or audience-participation programs, the format sometimes calls for a final contest to close the program. In other cases, where individual contestants are used, the format of some quiz programs provide for the program to be ended in the middle of one participant's competition. Similarly, in news programs, the idea of climax is not generally observed. News broadcasts follow the standard newspaper practice of putting important materials first, and less important material toward the end of the program. However, many newscasters do make an effort to provide some kind of final filling suited to listeners' and viewers' tastes. A novelty item, anecdote, maxim, or special climactic element, such as a short interview section or an editorial comment, is presented.

Attention has been given only to the final climax. In programs running more than thirty minutes, it is necessary to provide at least one middle climax. A program unit with climax values will precede the required break for station identification. In all, with programs longer than fifteen minutes, an important requirement is to try for an effect of building, while providing stronger and more interesting materials in the last half than in earlier parts of the program, to counteract inevitable lessening of listener and viewer interest. This occurs when the same sort of material or program, featuring the same entertainer, is presented for any appreciable length of time. Under climax, we not only have the requirement of providing a final climax unit and the middle climax unit in long programs, we also have, in relation to climax, the requirement of providing stronger and more attention-attracting materials for the latter half than in the first half of the program.

FINAL COMMENTS CONCERNING STRUCTURE

In summing up the discussion, three points should be emphasized. First, the well-constructed program must satisfy all seven of the requirements of effective structure, with the possible exception of climax in certain types of programs. Even here, if the idea of climax is abandoned, the structure of the program is made less effective.

Second, requirements apply to programs of all types, on radio, television, and cable. They have as much application for the disc jockey program, as for a sixty-minute program, to those of live music. Applications extend to straight talks presented on radio, television, and cable, as well as variety, or dramatic programs.

Third, the same seven requirements apply not only to broadcast and cable programs, but to every type of entertainment or communication. From the play presented on stage, nightclub circuit, and motion picture, to programs of oral communication. In every case, it's important to make attention as easy for listeners and viewers as possible. This applies to speakers, panel-discussion members, and interpersonal communication forum leaders, as well.

Chapter 10

The Purpose of a Broadcast and Cable Program

Whenever a radio, television, or cable program is put on the air, someone has to pay for it. Whoever pays those bills may be a sponsor, paying for the program and buying time for its broadcast or cablecast. It may be a nonprofit public service organization that gets time free from the station or cable facility, but still pays for costs of producing the program. Or, it may otherwise be the station or cable facility itself.

In every case, someone must pay the bills. The sponsor expects to get value received for money spent to put that program on the air. He isn't a philanthropist. He pays the costs of the program, or station or cable time because he wants to accomplish some specific objective. That objective is never to provide entertainment for listeners or viewers. Entertainment is the means to an end. It is the means used to induce people to listen or watch. He wants people to listen or watch and therefore hopes to influence those who do. Accordingly, his real purpose may be any number of things. If the program is sponsored, the purpose may be:

1) to sell some specific product. The "Ma Perkins" program was carried on network radio for 25 years to sell Oxydol detergent. The "Tennessee Ernie Ford" program was broadcast on television in the late '50s and early '60s to sell Ford automobiles. The Gillette Cavalcade of Sports program was broadcast to sell Gillette razors, blades, and other products produced by Gillette-owned companies, such as Toni Home Permanents and Papermate Pens. ABC Monday Nite Football is broadcast to sell Chevrolet products, General Motors parts, Lite Beer, and various motor oils.
2) to create goodwill for a company, or perhaps, an industry. This is called "institutional advertising." "Maverick" was sponsored by Kaiser Industries,

in part to sell aluminum foil and Jeep autos, and it also had the purpose of publicizing the building of goodwill through institutional advertising for Kaiser Industries in general.

In many cities, radio, television, and cable programs are sponsored by public utilities. For example, the local gas or electric company, or various telephone products companies, will sponsor programs. Frequently, these utilities have no particular merchandise to sell. However, they sponsor programs to create goodwill on the part of listeners and viewers, so when utilities ask for rate increases, there will not be too much criticism and opposition.

) to get votes for candidates for public office. Obviously, if a candidate buys time for a political broadcast during a political campaign, he has only one purpose in mind: to get people to vote for him on election day.

f the program is sustaining, but presented by an organization, the urpose may be:

) to induce to give money to a charity or to a service organization. If a program is presented by the local United Way, Red Cross, Tuburculosis Association, Heart Association, Muscular Dystrophy Association that produces the Jerry Lewis Muscular Dystrophy Associatin Telethon, Boy Scouts, or community self-help organizations, the purpose is to request money for support of these organizations.

) to influence attitudes of listeners and viewers in some desired direction. A number of programs are broadcast with the purpose of influencing listeners' and viewers' opinions, and ultimately, behavior. Church organizations present religious programs primarily to secure acceptance of their religious beliefs. Secondly, it is to induce listeners and viewers to attend some particular church and to increase giving and tithing efforts.

School systems and districts frequently use radio, television, and cable access channels to promote schools. They do this in order to make listeners and viewers more willing to vote for bond issues or to increase school tax levies. Nearly every organization to which a station or cable facility grants free time wishes to influence attitudes of those who listen or watch its programs.

to enlighten the listening or viewing public, to give information, or to raise cultural standards. In most cases, programs presented by nonprofit educational stations are intended to serve one or more of these objectives. We have programs on commercial stations, usually sustaining, whose real purpose is to inform the public, make them aware of the nature of important national problems, or to raise standards of music appreciation or drama.

There are not many such programs on commercial broadcasting stations, in terms of frequency, but programs of this type do exist. For nothing else, they meet stations' public-service requirements.

If the program is presented by a station with participating sponsorship, the purpose usually is:

1) to attract as large an audience as possible, so that as many people as possible may be exposed to commercial messages of various advertisers who buy commercial spot announcements. The purpose of advertisers who buy such announcements is to sell merchandise. However the program is put on the air and paid for by the broadcasting or cable facility, so it is that entity's purpose that must be analyzed. The larger the audience attracted, the more successful the program will be in selling merchandise for various buyers of announcements. Accordingly, it is much easier for the station or cable facility to solve the problem of getting advertisers to buy announcements on the program, or the station, for that matter, if the audience is as large as possible.
2) to build as large an audience as possible for that station or cable facility at a given hour. If a sufficiently large audience can be attracted, the station or cable outlet may be able to sell the time period when the program is presented, and possibly the entire program. Or, if it does not wish to sell time, building a large audience will result in a "carry-over" audience to the following commercially sponsored, or participating, program on that particular broadcast or cable outlet.

Chapter 11

Radio, Television, and Cable Program Ratings

When sponsors of network television programs spend money each week to present programs, they want to have some idea of the number of listeners and viewers reached by those programs. Even the local advertiser, buying time for a spot announcement on a local radio, television, or cable outlet, is interested in the size of audience expected to listen or watch his sales message. As a result, commercial research organizations have been developed with the broadcast and cable industry, providing figures at regular intervals which give approximations of the size of listening and viewing audiences for specific programs, or for an individual broadcast or cable offering, at some given hour during the day or night.

RATINGS

In general, three types of information are provided by rating services. First of all, a "rating" is a figure representing the percentage of all homes in a stipulated area equipped with receiving sets which, on a given date, had their receiving sets tuned to a particular program.

A rating for a television or cable program is based only on homes that have television receiving sets or cable hookups. For a radio program, it is based on all homes or autos having radio receiving sets. A national rating is a percentage of all receiving-set-equipped homes in a given group of cities. A local rating is a percentage based on receiving-set-equipped homes in a particular city, or metropolitan area, including that city and its immediate suburbs.

"Sets-in-use" is a figure showing the percentage of all receiving-set-equipped homes in a specified area that, during some specific period of time, were using their sets and had them tuned to any of several programs available in the area, during that time period.

In Oklahoma City, Oklahoma, for example, the sets-in-use figure for 3:00 P.M. on a given day would be the percentage of all homes, if one is considering television or cable homes located in the metropolitan Oklahoma City area, using their television or cable sets during that particular period. This includes those tuned to the available television or cable outlets.

A "share-of-audience" figure is one representing the percentage of all homes using sets at some given time, in a specified locality where sets were tuned to a particular station, network, or cable program.

In estimating ratings or sets-in-use figures, the base from which the percentage is computed identifies the total number of receiving-set-equipped homes in the area where the measurement is made. In estimating share-of-audience figures, the base is the number of homes where, at the time in question, a receiving set was tuned to some program. The sum of all share-of-audience figures for a specified time period of all stations serving a given community will always be 100 percent, whether or not half of all homes in the area had sets tuned in at the time in question, or whether or not only one home in ten was using its receiving set, or sets, at the time considered.

For purposes of illustration, let's refer to the metropolitan Oklahoma City area, roughly equivalent in area, to Oklahoma County, and including over 200,000 homes equipped with television or cable receivers. On a given evening a measurement was made, a check of a sample of 225,000 homes showed that half of all homes, or 50 percent, had their sets turned on between 7:00 P.M. and 7:30 P.M. The sets-in-use figure for television, then, for the thirty-minute period starting at 7:00 P.M. on that evening would be 50.0. Both ratings and sets-in-use figures are written and carried out to one decimal place. On the homes checked, 15 percent were tuned to the NBC affiliate, Channel 4. Consequently, the local Oklahoma City rating for the program on that date would be 15.0. The share-of-audience for that particular period of the evening in question would be the ratio of the rating to the entire sets-in-use figure. In effect, 15.0 is divided by 50.0, equaling 30 percent. This is for the share-of-audience for that NBC affiliate, from 7:00 to 7:30 P.M., on the hour in question.

TIME RATINGS

As a rule, rating information is provided for programs on either a national or local basis. Since advertisers buying time in given markets must make a choice among several outlets serving that market, adaptation of the rating methodology is applied to periods of time and individual outlets. For example, rating services compute the average rating for all programs carried on a given station or entity, on a specified day, during the three-hour period, averaged for five weekdays, Monday through Friday, in a given week. In markets where radio rating information is provided by one or more rating services, the ratings provided are for time periods on certain stations rather than for programs, as such.

In the same manner, total sets-in-use information for the same time periods is also provided. From the station's rating for a given hour, and sets-in-use figures for the entire community for that same hour, a share-of-audience figure for the station for that hour is computed. This is widely used in place of ratings for the station's advertising in trade magazines. This is done, primarily, because the share-of-audience figure is always considerably larger than that representing the station's rating for any given time period considered.

RESEARCH ORGANIZATIONS PROVIDING RATING INFORMATION

The practice of providing ratings for network programs began in 1930, with the organization of a company known as the Cooperative Analysis of Broadcasting. It was cooperative, since ownership was vested in a number of national advertisers and advertising agencies. The CAB used a telephone recall method of securing information.

Interviewers, in a number of different cities, called homes chosen at random from the telephone book and asked respondents to name programs where radio sets had been tuned in the preceding day. However, with other rating services available, using more reliable methods, the CAB went out of business in 1943.

For a number of years, another company, C. E. Hooper, maintained a virtual monopoly on the business of providing ratings for network programs. This company was organized in 1934. It took the major share of the rating business away from CAB and held dominance until after World War II. In spite of the competition from Pulse in years following 1941, the A. C. Nielsen Company bought out the Hooper organization, as far as national ratings were concerned, in 1949.

The original Hooper method was virtually identical with one used by a company named Trendex. Interviewers in a number of cities, each having an outlet for each national network, called homes chosen from the telephone book at random. They made a total of not less than 300 calls every fifteen minutes, between 8:00 A.M. and 10:30 P.M. They asked those who answered whether the family radio set was in use at the time the telephone rang. The Hooper method is called the Coincidental Telephone Method.

At the present time, there are a number of research concerns providing ratings on a national or local basis, or both, for television, cable, and radio. Four most important are Nielsen, Trendex, American Research Bureau, and Pulse, each using a different method of securing information.

Trendex

This rating concern used the same basic coincidental telephone method of securing information first developed by Hooper. Prior to 1958, it had interviewers in fifteen cities. Since early 1960, the number of cities has been increased. Trendex provides information on television programs only. Its ratings should be termed "multicity" rather than national, since information is secured only from television-equipped homes in cities where interviewers are at work. In all Trendex cities combined, a total of 1000 attempted telephone calls are made during every thirty-minute period on Saturdays and Sundays during daytime hours, and in weekday evenings up to 11:00 P.M.

During the daytime on weekdays, a smaller number of calls are made. Those for each period of the day, for all five days of the week, are combined to bring the total number of calls, which any rating is reported, up to the 1000 minimum figure. Information is secured during the first seven days of each month. A published report is made available to subscribers by the fifteenth of the month. No local ratings are provided, and no information is given concerning radio programs.

Pulse

Originally called The Pulse of New York, Pulse provides a variety of different ratings. There is a monthly rating of network radio programs based upon information secured in a number of markets. Local ratings of radio programs or time ratings for radio stations, are made, ranging from once a year to as often as twelve times yearly, in more than 200 markets. There is a national or multimarket rating for network and cable television programs based upon information secured in a number of markets. Local cable and television ratings range from twice a year to twelve times a year in several markets. A monthly average rating for all major syndicated filmed-for-television, HBO, CineMax, and Cable programs are offered by Pulse.

Pulse uses the roster method. Interviewers call at homes and secure information concerning exact hours the preceding day during times when radio or television sets are in use. They ask individuals to select the specific programs from a roster or listing, of all programs available during that preceding day. Local rating information and local ratings of syndicated films are based on a minimum of 300 home interviews for each day in the week. One-fourth are secured during each week in the month. National or multicity television and cable ratings of network programs are based on a minimum of 5000 interviews for each day of the week, secured in different metropolitan areas. The once-a-month network rating report for television and cable gives the individual rating for each of the metropolitan areas where the program is broadcast.

American Research Bureau

ARB provides television program ratings on a local basis across the nation. ARB also developed the use of ARBITRON. The regular ARB method of securing information is use of seven-day diaries, kept by the person for a seven-day period, with no financial payment made to those who keep the diaries. A different group of diary-keeping homes are used each month. Names of those asked to keep diaries are selected from telephone books at random, with preliminary arrangements made by telephone, so that nontelephone families are not included in the total sample.

The ARB national rating for television is based upon a sample of at lease 2000 homes, urban and rural, with information provided for the first seven days in the month. Local rating information is based on 250 to 500 completed diaries in each market. Approximately one-fourth of the total number of diaries will be completed for each week in the month. The total rating provided is an average of the ratings provided for four weeks.

Videodex

Videodex also uses the diary method. But there is a more or less permanent panel of diary keepers used, with payment made to diary-keeping homes. One-sixth of the panel is replaced each month, so that at the end of every six months, an entirely different panel is being used. National network program ratings are based upon panels from 9200 homes in a number of metropolitan areas, while local rating information is provided less frequently for approximately one to two hundred additional markets. Videodex national ratings are used less in the trade than those of Nielsen, Trendex, ARB, and Pulse. The major source of Videodex revenue comes from sale of local rating reports. Only television and cable are covered, with no information provided for radio.

Hooper

The Hooper organization, referred to previously, is still providing rating information, but solely on a local basis. All rights to national ratings were sold in 1949, to the company's leading competitor, A. C. Nielsen. Hooper provides local radio, television, and cable ratings to a number of markets. Local radio rating services are more important than television or cable counterpart services. In some markets, the coincidental telephone method is used, while in others, Hooper uses the diary method.

A. C. Nielsen

The A. C. Nielsen Company had built a wide reputation in market research before entering the program ratings field in 1941. The company is known for market research and program rating services. For its national ratings, Nielsen originally used a panel of 1050 television-equipped homes, scattered throughout the country, with rural as well as urban homes as a basis for national television ratings. A somewhat larger number, including 1050 television-equipped homes, is a basis for its national radio ratings today. In each home in a panel, a mechanical recording device, called an "audimeter," is attached to each set. It records, on moving tape, exact times when the set is turned on, and the channel, or frequency, to which it is tuned. The tapes run continuously over a fourteen-day period. At the conclusion, the tape in each audimeter is replaced and a new cycle of recording begins. National ratings for network and cable programs are released twice each month. Each report covers a two-week period. Separate reports are provided for network radio, television, and cable programs.

Nielsen's national ratings have been the standard in the industry since development of network television, and now cable. Nielsen also provides local ratings for television and cable, in over 300 markets. For local rating information, the diary method is used, kept during the first week of each month, in 200 to 5500 homes from each major metropolitan area.

ARBITRON

The most recent development in the rating field is a service of the aforementioned American Research Bureau. In each of several cities, a certain number of television-equipped homes are connected electronically with a central office. At intervals of ninety seconds during evening periods, and five minutes during daytime hours, an impulse is sent over the line from each television-equipped home to the central office, indicating whether or not the television set is on and to what station or access channel it is tuned.

This information is entered on IBM cards. At the end of each program period, it is possible to show average ratings for that time period, while reports are issued to subscribers on the following morning. The number of ratings-based homes, however, is decidedly small. There are several homes in a variety of areas hooked up with central offices. The major value of ARBITRON service is the speed with which information that can be supplied.

COMMENTS CONCERNING RATINGS

From time to time, critics in the industry launch criticisms of rating services. There is often a wide variation in ratings given for the same network or cable program, on the same date, by two national rating organizations. Consequently, questions are raised as to whether or not information provided by the services has any real lasting value. None of the rating services, however, can be expected to provide complete exact ratings for any program. The best expected results are approximate figures. In every case, the sample is relatively small. Statistically, an error of several ratings points one way or another is always possible. In every case, too, there is a real question as to whether the sample secured is completely representative of the entire population and totally random. All major methods used have characteristics that make complete or total accuracy impossible.

In the case of ARBITRON, the sample is extremely small, and there is a definite question as to whether homes chosen to make up the small sample are truly representative of all television-equipped homes in those cities that are used. If they are, the sample is limited at homes in well-defined, specified cities, with several outlets each. The listening and viewing situations in those cities will differ from that found in rural areas.

With national Nielsen ratings using a mechanical recorder, the sample is larger, and rural areas, as well as urban population outlays, are included. Nevertheless, the Nielsen audimeter reports only the television set was turned on. It offers no guarantee that anyone in the home is watching or listening. The coincidental telephone method used to provide national television ratings for Trendex, as well as the local ratings for Hooper, limits the sample to homes with telephones located in city areas. This is not truly an accurate cross-section of homes with television or cable sets throughout the entire country.

The Pulse method of interviewing listeners a full day after listening gives abundant opportunity for persons interviewed to forget precisely those programs tuned to twenty-four hours previously. Very popular and well-known programs and personalities are remembered, while programs less known are easily forgotten.

Use of diary methods have two serious shortcomings. First, 25 to 35 percent of all homes where diaries are kept fail to return diary forms. These nonrespondents may differ in listening or viewing habits from more interested ones that keep and return forms. Secondly, the diarykeeper may forget to fill out the diary form for a full day or even longer and then attempt to complete information on the basis of memory. If no rating is really completely accurate, and the methodology invalid, of what value are ratings?

They give ideas, proportionally of homes tuned to each program, on given dates. This provides an index of the proportional "likes and dislikes" of a given program, in relation to other programs. Although different rating systems supply varying figures representing ratings of the same program, on the same date, in the same area, *differences in method and methodology means various services are simply not measuring the same thing!* In most cases, ratings given for the same program are close to being the same. If any rating service reports a 15.0 rating for a given program, that is quite different from a 25.0. If any given rating company's rating figure for a program drops consistently, month after month, it is safe to assume the program is declining in popularity, and someone must do some soul-searching and patching very soon, or that program will go off the air.

PAST PROGRAM WITH HIGHEST RATINGS

In the heyday of radio, before television and cable took a major part of the home listening audience, the radio sets-in-use figure throughout winter months ranged from 40.0 to 50.0 during prime time evening hours, and 25.0 during the most attractive daytime hours. Since there were four radio networks until 1948, every radio station was a network affiliate.

The average rating of evening-sponsored network programs ranged from 10.0 to 15.0. The average rating of sponsored network radio programs during daytime hours was between 6.0 and 7.0. Today, the important sponsors are using television and cable, instead of radio. There are few important programs being broadcast on radio. This is a result of television and cable advertising competition, but also due, in part, because of lack of outstanding network or station programs on radio. The daytime sets-in-use figure for radio dropped to an average of 12.0 and the radio sets-in-use figure after 8:00 P.M. has fallen from 4.0 to 2.0. The result for the average radio programs today are ratings of 1.0 to 3.0 during the daytime, with lower ratings at night. The present total radio sets-in-use figure is divided in most communities, not between network outlets, but between 10 to 20 stations or outlets. This comes as a result of the tremendous increase in the number of radio stations currently on the air.

From 1930 to 1950, there were many radio programs consistently earning high ratings. Well-established radio network programs earned ratings of 25.0 or more during the winter months of one broadcasting season. The "Rudy Vallee Program," "Burns and Allen," "The Eddie Cantor Show," "Abbot and Costello," "Fred Allen," "Bull Durham Presents Will Rogers," "The Maxwell House Coffee Program," with Frank Morgan and Fanny Brice, "Lux Radio Theater," "Kay Kyser's Kollege," "Aldrich Family," "One Man's Family," "Mr. District Attorney," and Walter Winchell's "Sunday Evening News" were prime leaders.

High ratings were received by "The Bob Hope Show," "Jack Benny," "Fibber McGee and Molly," "The Red Skelton Show," and "Charlie McCarthy," each with winter seasons ratings of more than 35.0 for several years. Higher ratings were earned by two programs: "Amos n' Andy" had a 45.0 during the winter months during two seasons of 1931–32, while "Major Bowes' Amateur Hour" went above the 45.0 mark during the winter of 1935–36.

The highest rating ever reported for a single program on network radio, was 75.0, received by President Franklin Delano Roosevelt's "War Message to the Congress," after Pearl Harbor, December, 1941. This program was carried by all stations, and made available to outlets throughout the allied countries, as well as all American networks. Therefore, the rating and sets-in-use figures for the half-hour period of that broadcast were almost identical. The highest rating received by any program carried over one network was 68.0, received by the second Joe Louis title fight, carried on the ABC Radio Network.

HIGH RATED TELEVISION NETWORK PROGRAMMING

In the early days of network television, television novelty value was extremely high, with networks offering few outstanding programs each week, but some unusually high ratings, nevertheless, were recorded. In January, 1950, the "Milton Berle Program" received a 71.2 rating; while Arthur Godfrey's "Talent Scouts" had 50.9; Ed Sullivan's "Toast of the Town" received 47.6; and "The Goldbergs" had a rating of 44.4.

All of these were national Nielsen ratings. In January, 1950, not more than 10 to 15 percent of all American homes had television sets, and ratings were based solely on those homes with television sets, so high ratings represented only a small number of homes actually tuned to the programs.

Much larger actual audiences have been attracted to television network and cable programs in the early years of both. In December, 1953, for example, "I Love Lucy" had 58.7 for a rating, representing 15,000,000 homes. "Dragnet" had 54.8; the "Milton Berle Show" and Groucho Marx's "You Bet Your Life" each had a rating of 46.0.

During the 1959–60 season, the number of American homes with television sets stood at approximately 45,000,000, with 87 percent of all homes television set-equipped. The average sets-in-use figure for daytime hours during winter months ranged from 18.0 to 24.0. In prime time, the sets-in-use figure for winter months averaged close to 65.0. Consequently, the average network program, broadcast during daytime hours, with only three networks, had 6.0 to 7.0 ratings. The average sponsored television network program broadcast at night, in prime time, had a rating close to 20.0. The top-rated, once-a-week evening programs during the season were "Wagon Train" and "Gunsmoke," attaining 40.0 ratings or more. Occasional specials had ratings above that figure, but no program in recent years attained a rating close to the old "Milton Berle" rating of more than 70.0, received in 1950.

Chapter 12

Variations in Program Ratings: The Reasons

The rating received by any program now, national or local, depends on a wide variety of features and factors. Some of those affecting ratings of given broadcast or cable programs are discussed here.

NATION-WIDE VARIATIONS IN RATINGS OF NETWORK AND CABLE PROGRAMS

Despite minor variations resulting from rating techniques used, the national rating of a network or cable network program depends upon several factors.

Strength of Appeals Provided by the Program

Obviously, the stronger the program's appeals and strength for drawing large segments of the public, the greater the rating received. But strength of appeals is not the only factor considered. In some instances, it is less important than other factors, such as broadcasting at 4:00 in the morning! This is an extreme situation, but it illustrates strength of appeals, which alone, does not determine size of program ratings.

Extent Audiences Know or Are Familiar with the Program

A well-established program, where people are acquainted with the day and hour of broadcast, will receive a higher rating than a program of equally strong appeal values broadcast for the first time.

The disadvantage of a new program may be considerably offset by extensive publicity. Publicizing a change in the hour of broadcast will also help offset the handicap of a change in the broadcast time to an hour not previously associated with the program.

Hour of Broadcast

This is important because the sets-in-use figure varies considerably at different hours during the day. If appeals are strong, the program will generally get a higher rating if scheduled in prime time between 8:00 or 8:30 and 10:00 and 10:30 in the evening, or late at night. This is also true in case of television or cable programs.

Time of Year

The size of the sets-in-use figures for radio, television, and cable becomes very important. The sets-in-use figure, nationally, for radio, does not vary particularly at different times in the year.

In the case of television and cable, however, the onset of warm weather cuts down numbers of people who stay in the home and give attention to television or cable programming. The sets-in-use figure for evening television or cable viewing, as a whole, during the month of August, is two-thirds as great as the corresponding figure for January.

Prestige of the Network or Cable Facility

Prior to 1958, this was a highly important factor. The program carried on CBS and NBC would have a decidedly higher rating than that same program carried on ABC. Since fall, 1957, however, and since autumn, 1958, that situation has changed drastically.

Today, because of the Turner Broadcasting System for cable, and its various entertainment networks competing actively with the traditional national networks, there is no distinct advantage over competitors during evening hours. While cable facilities have made significant penetration in particular markets, the cable network prestige factor has had great effect on evening program ratings. Ratings have also been adversely affected by video cassette and disc recorders with playback capabilities.

Popularity of Lead-in Programs

If the program immediately preceding the one considered, on the same network or channel, has an unusually high rating, part of the audience for that preceding program will stay and "catch" the program under consideration.

This tends to make the program's rating higher than it normally would have been. On the other hand, if the preceding program has consistently low ratings, the chances of the program attaining a high rating, and maintaining its appeal values, are appreciably reduced.

Strength of Competition

If a program is broadcast at an hour when offerings of others are weak, the program will get a substantially higher rating than if forced to compete with extremely popular programs on other networks or cable outlets.

If a program is no more attractive than the competing program broadcast at the same hour on a rival outlet, the two programs will divide that part of the available audience enjoying that type of program. If high ratings are important, a network, station, or cable outlet shouldn't schedule variety against variety, western against western, quiz against quiz, or same against same. Finding a program type not being duplicated by the competition at the same hour is most desirable.

These are factors affecting the national rating of a network program. Local ratings, in various communities, for the same program, will also vary decisively. The same network shows might receive a 20.0 rating in one city and only 8.0 or 10.0 in another. This also holds for syndicated television film programs that receive very high ratings in one city, but a very low rating in another.

VARIATIONS IN RATINGS FOR THE SAME NETWORK OR CABLE PROGRAM IN DIFFERENT COMMUNITIES

The same principles discussed previously apply to radio, television, and cable, as well as network programs. Factors tending to make local ratings of a network television program higher in one community than in another include the number of outlets serving that market. Regardless of those outlets, the sets-in-use figure in each market, at a given hour, tends to remain reasonably constant and under reasonable restraints.

Assume that at the time a given program is on the air the sets-in-use is figure 60.0. This means that in each of two communities under consideration, 60 percent of all outlet-equipped homes have sets tuned to some program. In another community served by a two-station market, without outside service, that station gets half of 60.0, or a rating of 30.0. In a city with four outlets, however, the share of each station would be only one fourth of the 60.0, sets-in-use figure. The rating average would be only 15.0. This means that the greater the number of outlets effectively serving a given market, then the local rating, of any given network program, will be lower. The smaller the number of outlets, the higher that local rating of the same program.

Prestige of the Station Carrying the Program

Just as one network or cable company may enjoy a greater degree of prestige than competitors, there is a substantial variation in the degree of prestige enjoyed by several stations or cable channels in that community. One may, because of superior local programming or affiliation with a network, have better or more popular programs than the others. This means that one station, or outlet, can build a habit of listening, or viewing, in a larger proportion of homes in that community, than other outlets. In doing so, it can maintain a position of advantage over others in that community or market.

If a network program is carried on a high-prestige station in the community, it will have a higher rating than it would have received, with the same strength of appeals, if carried on a low-prestige status station in the same city. Station or cable prestige, then, is developed by carrying consistently good and popular entertaining programs, with consistent promotion of those programs on the air. It is also developed by use of strong local personalities appearing regularly on that station or cable facility.

Amount of Local Promotion Given the Program Considered

If the network program is given extensive publicity and advance advertising in one city, while no advance publicity is accorded it in a second community, the use of that extensive advance promotion will bring the program a higher rating than would be the case if little or no advance advertising is given.

Since there is often considerable variation in amount of program promotion given a network program in different cities, there will be variation in the local ratings in different cities, as a direct result of that variation in amounts of publicity.

Section of the Country Where a City Is Located Will Affect Local Ratings

Considering the same network television program, the rating received by that program will be higher in northern parts of the United States than in the South. This is for exactly the same reason that ratings are higher during winter months than during summer months. In the South and in the sun-belt area, higher-than-average temperatures at any given time in the year mean people spend more time outdoors than in cities such as Minneapolis, Milwaukee, Chicago, and Detroit. With more people outdoors, the use of television sets is lower. Programs will have lower ratings in warm, moderate southern climates and the sun-belt, than in the north.

There is also an east-west factor affecting local ratings of network or cable programs, as a result of time zones. A program broadcast at 8:00 in the evening in the Eastern time zone will be heard at 7:00 in the Central time area, 6:00 in the Rocky Mountains, and 5:00 on the West Coast. The sets-in-use figure is appreciably lower at 5:00 than at 8:00, during prime time.

Ratings of programs broadcast at 8:00, Eastern time, are usually lower in the Central time states than in the East. Most television and cable network programs are taped and rebroadcast at a later hour than the hour seen in the East or on the West Coast. Sometimes this change in time helps local ratings in Western and Mountain time cities, while at other times, it is a definite handicap.

The local ratings received by any given network program in the 100 to 200 cities carried will vary widely from one city to the next, even though the appeals provided are identical in all cities. The appeals provided by a program are certainly not the only factor influencing either national or local ratings. A number of important factors must be taken into account.

Syndicated Television Programs

The situation with syndicated television programs is more complicated than national network or cable programs. There are exceptions, but the ordinary network program seen on the same day and hour in Cleveland, Pittsburgh, or even New York, and allowing for time zones at the same time, if not the same clock hour, will also be seen in Chicago and St. Louis.

But a syndicated film has no regular time. It is only by "accident" that the same episode is broadcast on the same day, date, or same hour in any two major cities. In one city, the program may be scheduled at 11:00 or 11:30 at night; in another, on a different day of the week, perhaps at 6:30 in the evening; while still in a third, it might be a period without network service at 9:00 in the evening. The time factor, then, is highly important in producing variations in local ratings. Local ratings for syndicated programs vary even more widely than local ratings received by television network programs.

Locally Originated Programs

Programs on television, cable, or radio, in a single city, by an originating station, such as local news programs, radio disc jockey programs, and others will always highlight local factors. Appeals, publicity, hour of broadcast, time of year, habit of viewing, popularity of the program preceding, and strength of competing programs on other stations or access channels, all apply with sole exception to the prestige of the network or cable entity, however, no network or cable operation is involved. In the case of cable, it is local-access channel orientation that governs this situation.

Ratings of radio programs will be consistently lower than those of television or cable programs at the same hour. This is due to the fact that the sets-in-use figure for radio is lower than that for television at practically every hour of the day, except before 8:00 or 9:00 in the morning, every day of the week, and 5:00 and 6:00 in the afternoons. These periods are called "drive times."

There is a smaller total audience from the general audience of a given station that can be drawn. During daytime, sets-in-use figures for radio will be greater than for television and cable prior to noon, or 1:00, and at night during prime time. The sets-in-use figure for television is likely to be 50.0 or above, compared with a sets-in-use figure for radio of only 6.0 to 8.0

As regards daytime ratings, the number of radio stations serving any given market is usually three or four times as great as the number of television or cable outlets. If there are ten radio stations, the average rating will be only one-tenth of the radio sets-in-use figure, while with television and cable access channels, the average rating will be that proportion of the television and cable sets-in-use figure at the same hour, which makes a big difference.

Chapter 13

Appeals in Programs

In discussing "appeals," one finds some programs are more effective in attracting listeners and viewers and holding attention of those members of the audience they do attract. This is a result of differences in programs. Programs usually attractive to audiences, yet effective in holding their attention contain elements which have compelling values. These elements make the individual want to listen or watch, and can be referred to as "satisfactions," provided by the program. Psychologists refer to them as "motivations," or motivating factors, included in the program. These factors offer a motive to the listener or viewer. However, broadcast professionals in the industry refer to these factors as audience appeals, or simply appeals, provided by the program.

The term *appeals* is used totally throughout the industry, in the same sense the word is used in this book. These factors, recognized by broadcast and cable professionals, are referred to frequently with respect to material provided by programs. Despite what they are called, planners of programs recognize program attractiveness can be increased by introduction of various elements, although no systematic effort has been made to define these elements, or group them under any single term such as appeals.[1]

Various types of appeals are provided in radio, television, and cable programs. The more important appeals used can be grouped under seven major headings:

1) Conflict or Competition
2) Sex Appeal
3) Information
4) Comedy
5) Human Interest
6) Emotional Stimulation
7) Importance

Very few programs offered on radio, television, or cable provide all seven appeals. It isn't necessary; when a program provides two, three, or four, they will attract audiences and hold attention. Use of additional appeals may be helpful but are not essential. Any program providing as many as three is certain of being a successful program in attracting large audiences.

Seven major appeals should be considered in detail, with attention to forms certain appeals take, and the manner introduced into programs.

CONFLICT OR COMPETITION

This is the most basic of all appeals. It is difficult for a program to be effective without providing some element of conflict. Audiences are interested in situations that involve conflict. A one-sided struggle creates little interest. Whenever thinking of conflict appeal, think of it as including some element of uncertainty or suspense, as well as being an ordinary struggle.

Types of conflict appeal used in broadcasting and cable include these additional factors.

Struggles Involving Danger to Life or Physical Safety

These are provided primarily in thriller dramatizations where the hero, with whom the listener or viewer normally "identifies," is in danger of being killed or threatened with torture or suffering. Some actuality broadcasts similarly show individuals in physical danger; however, the physical danger type of conflict is provided in fictional dramatic thriller dramas.

Physical Conflict, without Danger to Life

This is the type of conflict provided in broadcasts of sports events, including boxing, wrestling, football, and soccer, where there is a considerable degree of body conflict.

Conflict in Noncontact Sports Events

This appeal is popular in broadcasts of baseball or basketball games. Here the physical prowess of the hero, or members of the team, will provide the factor of conflict. Other types of sports events such as bowling, horse racing, and polo are included in this same category.

Conflict Based on Threats to Reputation or Business Success

This form is widely found in nonthriller dramatic programs. As a rule, the hero finds himself in a situation involving danger, but not physical danger. For many, danger to reputation or threats to success may be no less strong as an appeal than that related to the hero's physical safety.

Conflict in Love

Dramatizations based on "triangle situations" provide high degrees of conflict appeal. Even more common in radio, television, and cable is use of the love conflict between a young man and woman. Their ultimate happiness is threatened by misunderstandings or machinations of an "evil" third person, who attempts to break up their romance and happiness.

Problem-solving Conflict

This type of conflict is provided by quiz programs. The appeal is prevalent whether the viewer is vicariously suffering with the contestant who must answer a difficult question, or whether the viewer, himself, attempts to arrive at the proper solution of the problem. Detective-type dramas make use of problem-solving conflict, in situations where the identity of the "villain" is not revealed until the very end of the program. The viewer, consequently, is encouraged to attempt to select the "villain" before his identity is revealed on the program.[2]

Conflict of Ideas

This conflict is provided in discussions of controversial issues on forum programs, broadcasts by political candidates, and others, with propaganda motives. It is a basic element in one-sided presentations by news commentators, within the dimension and discussion of controversial issues.

Human Internal Conflict

Another way that formidable conflict is effectively introduced is by development in dramatic programs specifically, and in other program types or situations where the hero or other person, with whom the audience will identify, is forced to make a difficult decision. In many cases, entire dramatic programs are built around this type of problem. The viewer, as a result, suffers along with the hero.

These are not the only types of conflict possible. They are listed to give dimension to the variety of conflict included in programs. Any use of comedy involves use of conflict. Comedy is, normally, at the expense of someone upon whom a joke is played or who becomes the butt of a prank, or who is made to look ridiculous. The term *slapstick* suggests a way comedy may become physical. For this book, however, conflict provided in comedy is considered a part of comedy as an appeal, while not involving conflict, as such.

COMEDY APPEAL

Comedy can be a highly effective appeal. Most listen or watch programs for entertainment because they like to laugh. Some programs offer comedy as the strongest appeal. In a large number of programs, comedy is used, and made to be, a reinforcing appeal.

As used in radio, television, and cable, comedy can be provided through lines or statements that are funny. It may also be provided by use of characterization, or through situations. Sometimes all three forms are used in a single program. For general purposes of classification, it is best to present comedy appeal within six major classifications.

Broad Slapstick Comedy

This type of comedy is provided in some of the "Three Stooges Programs," but is best illustrated perhaps in old films featuring Laurel and Hardy, Abbott and Costello, or the Keystone Kops. As a rule, clever lines are not involved. Comedy comes from placing less-than-highly intelligent characters in ridiculous situations or having those same characters engage in dialogue, often through exchanges of insults.

Invariably, characters are overdrawn, dialogue in the situations are implausible, and part of the comedy will depend on physical violence, in a mild form. The term "slapstick" describes the type of comedy presented.

Broad and Exaggerated Farce Situation Comedy

This form of comedy is provided in dramatic programs, where exaggerated characters with broad and extreme situations are used. The form is represented by such programs as "The Jeffersons," "Movin' on Up," or "The Dukes of Hazard."

Exaggeration in character and situation will contribute to comedy values, with no effort made to make either characters or situations plausible.

Realistic Situation Comedy

This comedy form is used primarily in comedy-type dramatic programs; however, characters are realistic. The comedy values arise from putting ordinary people in comedy situations that might happen to anyone. Many, but not all, family-type comedy dramas depend on this type of comedy as a major appeal. Illustrations are "Eight is Enough," "Laverne and Shirley," "The Love Boat," and episodes of the "Barney Miller Show."

Gag Comedy, Using Characters

Comedy is based chiefly on lines or anecdotes told by a single person, or lines in dialogue presented by a comedy team. There is no plot, while the comedy comes out of what is said. The person who presents the comedy lines will appear as a comedy-type personality, for example, as a "hillbilly" on "Hee-Haw." Under the circumstances, a very broad type of comedy is used.

Gag Comedy, by Straight Comedians

The entire emphasis is on use of clever or laugh-provoking lines. There is no plot or situation, because the comedian appears as an ordinary person. Usually a monologuist, but sometimes a character in a dramatic program, either can be used in gag comedy. Bob Hope or Johnny Carson provide outstanding illustrations of gag comedy.

Sophisticated Comedy

There is relatively little of this type comedy or humor found today on radio, television, or cable. The best explanation of this type was provided by Will Rogers or Fred Allen. It usually deals with public events of current importance. In talk or monologue form, it does not involve situation or special comedy characterization and is not dependent on dramatic situation.

There are types of comedy that do not fall within any of the types named. But these types give an idea of the wide variety of comedy appeal. Programs such as comedy variety and comedy drama are programs where comedy is the one outstanding appeal. However, in many other dramatic programs, a character is introduced to provide comic relief, such as the Festus character in "Gunsmoke," or where an effort is made to provide comedy as a secondary, or supporting type appeal. Some types of audience participation programs give heavy emphasis to comedy, either in interviews or stunts performed by participants. In many quiz programs, a comedian is used as quiz master.

In vaudeville variety or country/western music programs, part of the total time is used to present a featured comedian who is featured regularly on that program or a special guest.

SEX APPEAL

This is a type of appeal with which every radio, television, or cable audience is familiar. It often takes the special form of love interest in a program, making the sex appeal label cover a more wide variety of material than might be expected. It is present in radio, television, or cable programs in many ways.

Physically Attractive Men and Women, as Principal Characters or in a Minor Role

Sex appeal is also provided by use of physically attractive men and women. As a rule, the hero of a dramatic program satisfies this requirement, but sex appeal may also be found in the personality of a master of ceremonies, panelist on a panel show, or even a featured comedian. The personality providing the appeal may be either male or female. The important element is the degree of attractiveness and warmth of personality.

A considerable part of sex appeal value from a program personality is a result of that personality's speaking voice. It may in no sense be "sexy," but if it is warm, pleasant, friendly, or inviting, it will contribute to sex appeal value for that program.

Use of Love Stories

Programs dramatizing "boy-meets-girl" situations have strong sex appeal values. A degree of sex appeal is also provided emphasizing love angles with older, married couples. Major sex values in triangle situations, common in daytime serial dramas, also meets this requirement. Love stories can be told in audience participation programs, when contestants are asked to tell how they first became acquainted or how one marriage partner proposed to the other.

All references to marital love, wedding ceremonies, or being engaged have sex-appeal values. The appeal is strongest when presented as a dramatization of the relationships and misunderstandings of young couples, who at the close of the dramatization, become engaged or are actually married.

Music with Lyrics Dealing with Love or Courtship

Ninety-five percent of all current popular music, and a considerable proportion of musical standards, have vocals that either express love of the vocalist for a member of the opposite sex or deal with unrequited love. Naturally, such musical numbers have strong sex-appeal value. Even if the music is handled entirely on an instrumental basis, with lyrics not used, a large proportion of listeners "fill in" the lyrics in their own minds, if that number is one with which a large proportion of listeners are familiar.

Old, familiar music having a love theme is strong in sex appeal values for middle-aged or older listeners. An example is the popularity of the German soldier's song "Lili Marlene," sung by allied armies at the close of World War II. For those who were high school or college age and falling in love for the first time when the music was first popular, the song presents strong sex appeal values. This accounts for the tremendous audience response to "oldies but goodies" programming by a number of radio stations today. Also, the post-World War II baby boom has added a great deal of emphasis upon numbers of young men and women now in adult age categories today, who can remember when that music was first released or popular.

Music Generally, Regardless of Theme or Lyrics

All music has some sex appeal value. Even rhythmic beating of drums or tom-toms has sex-appeal connotations. Values of classical music are less than those of other types, but even here, some degree of sex appeal is provided.

HUMAN INTEREST

This is a moderately strong appeal in many programs and extremely strong in others. In general, the appeal arises from interest in people. It is concentrated in ways people behave, from their problems to things that interest them. It may be provided in a variety of different ways.

Ordinary People

In audience participation programs, participants who appear are ordinary people, like those who live next door. In dramatic programs, human interest values are made stronger if the characters presented are ordinary people. The quality of personality is particularly emphasized, even though exaggerated, in most country/western variety programs. In the "Barney Miller Show," pains were taken to make Barney a real person through character emphasis, possessing a unique understanding of people and their personality traits. The character of Barney Miller has many identifiable human qualities.

The greater the degree an entertainer avoids stiffness and formality, while permitting his character to make mistakes, the more the audience will see him as the person living next door. Consequently, human interest value grows stronger through these very clear personality characteristics.

Presenting Problems of Ordinary People

Some audience participation programs are deliberately built around ideas dealing with ordinary people, others in the way people react to problem situations. We like to know what bothers people and how they feel about things that the character on-camera thinks about.

Use of Ordinary Situations

Although human interest is a keen interest in people, the value of human interest appeal can be heightened by placing people in ordinary, everyday situations. For example, in a comedy-dramatic program, the plot may revolve around commonplace family situations. Problems arise when Junior brings home a report card showing a low grade in English, or family turmoil is created when a teenage daughter goes to her first formal dance, or conflict arises when the husband-wife disagree with respect to plans for a family vacation.

Point Ordinary Things

One device used to heighten human interest values is having people do ordinary things, in the way of stage business. The leading character can't find a lighter to light his cigarette, making the cigarette lighter work, difficulty tying his necktie, or bumping into a piece of furniture. All these difficulties contribute to human interest values.

EMOTIONAL STIMULATION

This is closely related to human interest, but goes beyond it in scope. We're interested in people, as well as their problems and troubles. When problems are serious enough to arouse a feeling of sympathy for them, in their difficulties, then, not only has the appeal of human interest prevailed, but it also becomes part of emotional stimulation. Sympathy is the major form of emotional stimulation provided.

Emotional stimulation can develop from introduction of babies, young children, or the elderly in a program. It may take the form of an appeal to patriotism, or loyalty to highly regarded institutions or strongly held beliefs. The introduction of materials relating to religion, including a short prayer or the singing of a hymn in the program, heighten this appeal. It provides strong emotional stimulation for audiences. All references to God, home, or mother themes are sources of emotional stimulation. What should not be classed under emotional stimulation are less desirable emotions, such as hatred and fear. Both of these are considered as adjuncts to the appeal of conflict. We reserve the appeal of emotional stimulation to emotions such as sympathy for others, nostalgia, religious interests, love for parents or children, interests related to the very young children, and the elderly, as well as appeals to various loyalties, including patriotism.

INFORMATION

Information can be an extremely strong appeal, or lacking entirely. The variation depends upon the degree of program subject matter and how it affects vital interests of the audience. If the subject has life or death significance, information will be a powerful appeal. If the subject does not concern the audience, information will be negligible or nonexistent.

In between, there are stages of appeal value for information. We are interested, for example, in news programs dealing with happenings that will affect us, relating to people we know, or places we are acquainted with. We may be strongly interested in purely local news, relating to our own community, but less interested in news of national significance, unless the newscaster shows how events in Washington, San Francisco, New York, Moscow, or Peking have direct bearing on our own well-being. There is only casual interest in events taking place in foreign countries, unless those events involve people with whom we have some prior acquaintance.

"Odds and ends" information has moderate appeal value, and nothing more. The same is true with information about strange places, or unusual happenings. General information, in organized form, might be educational, but has appeal value for exceedingly small audiences. People listen to radio, or watch television or cable outlets, to be entertained, not educated.

IMPORTANCE

The final appeal influences programs in three ways. First, the program may deal with subject matter inportant to the listener or viewer, but things of vital importance may be of little interest to individuals. Secondly, the program may offer name value and feature well-known entertainers with outstanding reputations as comedians, actors, or vocalists. It may present those important in politics, business, or education. Thirdly, an effect of importance can be provided through an elaborate production, with a large cast, orchestra, or elaborate scene design. Importance and information are rarely used to carry a program alone. As a rule, both serve as secondary appeals, reinforcing other possible appeals in programs.

FACTORS THAT STRENGTHEN APPEALS

Every program, in order to attract listeners or viewers, while holding audiences' attention, must offer two or three of the seven basic appeals. It isn't necessary for the program to provide all seven. Very few programs will offer all seven. In successful programs, two, three, or four may be offered strongly. There may be another one, two, or three provided as secondary appeals to help in holding interest, but not strong enough to be major reasons the program attracts an audience.

A discussion of appeals is not complete without attention given to three additional factors serving to supplement and enhance the appeals provided. These factors heighten the overall appeal value of the program. Involvement, plausibility of program materials, and the novelty of concept and material within the program should be discussed in detail.

A Program Is Stronger and Appeals More Effective if the Audience Has a Sense of Being Involved in the Program

Involvement is sometimes referred to as participation. It takes the form of empathy, or a tendency to identify with a character in a dramatization. It can be with a contestant in an audience-participation quiz program. Sometimes necessary involvement is provided by the individual's feeling he is affected by materials presented. In other program circumstances, involvement comes from the listener or viewer being acquainted with an individual, place, or situation important in the program. In considerable degree, strength of any of the seven appeals depends upon the degree a listener or viewer is involved in what is being presented on the program.

Involvement in Conflict Situations

Conflict, on its own, may be a strong appeal. But strength of conflict is tremendously heightened when the conflict presented involves the listener or viewer. In dramatic or audience participation programs, where listeners or viewers are made to identify with one or more of the participants, the degree of involvement will be high and conflict appeal will be heightened and made much stronger.

In programs of sports events, strength of conflict appeal can be heightened if the audience has a direct interest in one of the participants. For example, one of the teams in a football game represents a school the audience can identify with, or an HBO audience selects one contestant in a boxing match or a horse-race and consequently "pulls" for that contestant to win; then conflict is made persuasive in content.

If the conflict involves ideas or principles, and the subject matter is one the audience shows strong feelings for, there is a feeling of involvement. Consequently, response to conflict appeal is made much stronger. If a dramatic show presents characters so implausible, unpleasant, or unsympathetic that the audience will not allow itself to identify with the leading character, then conflict appeal is tremendously weakened. The same is true in a program of sports events. If the listener or viewer has no particular interest in either participant, and doesn't care who wins, the appeal process will break down completely.

Involvement in Sex Appeal Situations

Involvement is essential for targeted age groups. In a love story, sex appeal is strong only if the listener or viewer can identify with one of the characters involved in the love story. Identification seems easier for women than men. In most cases, sex appeal is much stronger, as an appeal, for women than for men. A love song popular 20 years ago has tremendously greater sex appeal value for people who knew, and were affected by, the song when it was first popular than by people unfamiliar with the melody.

Involvement in Human Interest or Emotional Stimulation

If human interest is to be an effective appeal, the character, entertainer, or audience participant must arouse human interest and be like someone the audience knows to be a real person. The greater the resemblance of the character to a person who lives next door, the stronger the human interest appeal.

As a result, there are strong or very strong human interest and appeal values provided in dramatic stories involving ordinary people in reasonable daily situations. But human interest appeal is extremely weak or nonexistent in dramatizations involving "upper crust" British society. The same principle applies in other programs, as well. Similarly, in case of emotional stimulation, the introduction of a baby or a small child has much stronger stimulation value for women who have had children than for childless younger listeners or viewers. There is interest for what the audience knows about children. Superman has little human interest or emotional stimulation appeal value for some audiences. The character simply doesn't fit the test of being a real person.

Involvement in Information Appeal

While there is limited appeal value in "odds and ends" information, it is greatest when it immediately affects the listener or viewer, and their immediate interests. To have strong information appeal value, a news item must deal with a person known to the audience. With a person who is a national figure, discussed in earlier news stories, or a situation, problem, or place, known to the audience, the audience, therefore, is already interested, and information appeal will be significantly strengthened.

A program giving information about wheat farming may be of considerable interest value for those engaged in wheat farming but offers little involvement to average city dwellers. Consequently, it provides no special degree of information appeal for the mass audience. The more the individual listener or viewer is concerned with information, the more information appeal will be heightened in a program. Accordingly, having information deal with people, places, situations, or problems, that a specific audience knows, and is interested in, the greater the involvement and effectiveness information will be, as an intrinsic appeal, in a program or program series.

Involvement in Importance Appeal

Elaborate productions are exceptions to the general rule for this appeal. But other types of importance appeal, however, such as use of names, and the consideration of important situations or problems, will depend on specific audience involvement. An entertainment name is important only to those familiar with that name, and importance in the entertainment industry, for that particular name entertainer, rests with basic audience involvement.

For example, a name important in government, business, or public life is similarly endowed with importance appeal value to those who know that person, or something about the importance provided in subject matter. Importance appeal is provided only for those aware of the importance of the subject, and particularly in situations where the subject is of special importance to specific individuals.

Therefore, it is evident that effectiveness of every appeal is dependent upon the degree the individual has of becoming involved. It is a sense of being affected and interested in material presented, or involved with the entertainers and others taking part in the program. Involvement varies among different individuals with differing backgrounds and existing interests. As a result, the effectiveness of any given appeal differs among those with different interests and backgrounds.

Accordingly, materials presented must be plausible and believable in most types of programs. There are exceptions, however, such as cartoon comedies, overdrawn situation comedy dramatic shows, and thriller dramatic programs for children. In the majority of programs, it is important audiences feel characters and situations are believable. Characters who give an impression of being real people, behaving in actual manners everyday people would behave in similar situations, and presented in believable surroundings, will be most successful.

Radio, television, and cable drama is make-believe, but we demand that those things taking place on the air be reasonable and could happen. Characters that appeal and conform to our ideas of what characters should be like in real life will be successful. We won't believe it if a character in a dramatization apparently has supernatural powers. We won't become involved if he behaves in a manner not conducive to an ordinary person under similar conditions and circumstances. Believability is spoiled if a character, in an adult nighttime serial such as "Dallas" or "Dynasty," talks like a "grease monkey," or if in the background scene representing a metropolitan street in Dallas, Texas, or New York City, the camera shows a range of rugged mountains, or if a New York subway conductor is presented as living in an apartment of the $10,000 a year rental type.

We want to believe materials given in the program are real and true. The total attractiveness of the program, then, is substantially weakened if people, settings, and situations presented in the program aren't believable. The examples given are extreme, although such obvious implausibilities occur in many such programs. The most frequent error in believability results from actions on the part of the character not fitting the person portrayed. Actions in any way out of the ordinary must be given sufficient motivation, in advance, while some actions simply won't happen, regardless of motivation. Appeals presented in a program will be seriously weakened, or countered completely, if characters, situations, and actions of characters fall short of being plausible and believable.

Finally, the effectiveness of a program will be affected by degree of freshness, newness, originality, and novelty provided. The concept of freshness is difficult to draw complete conclusions or answers. We know programs and personalities that have been continually successful in attracting audiences over periods of many years. At the same time, every program has a tendency to wear out after newness, freshness, and novelty have worn off. Some programs highly successful during the first year or two on the air have fallen off rapidly in attractiveness by the end of their third or fourth season. This tremendously rapid turnover in evening network television and cable programs is sufficient evidence of the wearing out of the most popular programs. Furthermore, some programs have skyrocketed into popularity within a period of a few weeks or months because they offer something new and quite different from other programs available.

As newness or freshness wears off, the program will decline in attractiveness, even though the appeals that are provided remain significantly unchanged. Further, in comparing two programs that offer appeals of equal strength, the program with a higher degree of difference, or one using more original treatment will likely have stronger overall attractiveness than other competing programs. Therefore, different quality is a decided asset to any program attractive to audiences.

NOTES

1. In discussing motivations or attitude perspectives, one must be aware of an overview of the entire area of attitudes and attitude change, as well as specifics in the area of learning theory, social judgement, and cognitive balancing. Attitudes are typically defined as predispositions to respond in a particular way toward a specified class of objects, according to J.M. Rosenberg, in "The Experimental Investigation of a Value Theory," *Journal of Abnormal Social Psychology*, 52, (1953) p. 12. As predispositions they are not directly observable or measurable. Instead, they are inferred from the way we react to particular stimuli. The types of response that are commonly used as indices of attitudes fall into three major categories: cognitive, affective, and behavioral.

Within the current state of attitude theory, there is no one available model to be judged clearly preferable to all others; the researchers do not seem to employ a completely uniform vocabulary or uniform set of concepts. Each researcher pursues a particular and special way of formality and formally casting up some of the general issues that have been raised in some particular work. In this vital area of research, one must look for continuities and for inconsistencies in order to appraise for himself the ragged lines of progress and gaps of ignorance that characterize current efforts in attitude theory and research. By such active participation, hopefully, someone may contribute to mass media research in solving the enormous problems posed by attitudes and their change, and as they relate to what we term motivations. A list of references is provided to show directions of some of the research within this area.

2. See the following references for clarification as to how problem-solving conflict relates to attitudes, and attitude change.

Abelson, R.P., Rosenberg, J. M. "Symbolic Psycho-Logic: A Model of Attitudinal Cognition." *Behavioral Sciences* 3 (1958): 1–13.

Adorno, T.W.; Frenkel-Brunswik, Elsie.; Levinson, D.J.; Sanford, R.N. *The Authoritarian Personality*. New York: Harper Bros., 1950.

Axelrod, J. "The Relationship of Mood and of Mood Shift to Attitude." *Technical Report No. 5 to the Office of Naval Research*. Rochester: University of Rochester, (1959): 5.

Burdick, M.A. Burns, A.J. "A Test of 'Strain toward Symmetry' Theories." *Journal of Abnormal Social Psychology* 51 (1958): 367–70.

Carlson, E.R. "Attitude Change Through Modification of Attitude Structure." *Journal of Abnormal Social Psychology* 52 (1956): 256–61.

Cartwright, D., Harary, F. "Structural Balance: A Generalization of Heider's Theory." *Psychology Review* 63 (1956): 277–93.

Festinger, L., *A Theory of Cognitive Dissonance*. Evanston: Row-Peterson, 1957.

Green, B.F., "Attitude Measurement." In *Handbook of Social Psychology*, edited by G. Lindzey et al., vol. 3. Cambridge: Addison-Wesley, 1954, 1021–61.

Harding, James B.; Kutner, H.; Proshansky, L.; Chein, L. "Prejudice and Ethnic Relations." In *Handbook of Social Psychology*, edited by G. Lindzey et al., vol. 2. Cambridge: Addison-Wesley, 1954, 335–69.

Hartley, E.L. *Problems in Prejudice*. New York: King's Crown Press, 1946.

Heider, F., "Attitudes and Cognitive Organization." *Journal of Abnormal Social Psychology* 21 (1946): 107–12.

Heider, F., *The Psychology of Interpersonal Relations*. New York: Wiley, 1958.

Horowitz, M.W.; Lyons, John.; Perlmutter, H.V. "Induction of Forces in Discussion Groups." *Human Relations Quarterly* 4. (1951): 57–76.

Hovland, Carl.; Harvey, O.J.; Sherif, M. "Assimilation and Contrast Effects in Reactions to Communication and Attitude Change." *Journal of Abnormal Social Psychology* 55 (1957): 217–32.

Hovland, Carl. "Effects of Mass Media of Communication." In *Handbook of Social Psychology*, edited by G. Lindzey et al., vol. 2. Cambridge: Addison-Wesley, 1954, 1063–03.

Jordan, N. "Behavioral Forces that are a Function of Attitudes and of Cognitive Organization." *Human Relations Quarterly* 6 (1953): 273–87.

Katz, D., Scotland, E. "A Preliminary Statement to a Theory of Attitude Structire and Change." In *Psychology: A Study of a Science*, edited by Samuel Kock et al., vol. 3. New York: McGraw-Hill, 1959, 423–475.

Katz, D., Braley, R.L. "Racial Stereotypes of one-hundred College Students." *Journal of Abnormal Social Psychology* 28 (1933): 280-90.

Lazarus, R.S. McCleary, Robert A. "Automatic Discrimination Without Awareness: A Study of Subception," *Psycological Review* 58 (1951): pp. 113-22.

Lund, Frank H. "The Psychology of Belief," *Journal of Abnormal Social Psychology* 20 (1925): 63-81.

McGuire, William J. "The Nature of Attitudes and Attitude Change." In *Handbook of Social Psychology*, edited by G. Lindzey et al., vol. 3. Cambridge: Addison-Wesley, 1954, 440-93.

Murray, H.A., Morgan, Christiana D. "A Clinical Study of Sentiments." *General Psychology Monographs* 32 (1945): pp. 310-12.

Murphy, George; Murphy, Lois; Newcomb, T.M. *Experimental Psychology*. New York: Harper Bros., 1937.

Nowlis, Victor. "Some Studies of the Influence of Films on Mood and Attitude." *Technical Report No. 7 to the Office of Naval Research*. Rochester: University of Rochester, (1960): p. 5.

Osgood, C.E.; Sugi, G.J.; Tannenbaum, P.H. *The Measurement of Meaning*. Urbana: University of Illinois Press, 1958.

Peak, Helen. "The Effects of Aroused Motivation on Attitudes," *Technical Report No. 8 to the Office of Naval Research*. Ann Arbor: University of Michigan, (1959): pp. 52-58.

Peak, Helen. "Psychological Structure and Psychological Activity." *Psychological Review* 65 (1958): 325-47.

Rosenberg, J.M. "The Experimental Investigation of a Value Theory." *Journal of Abnormal Social Psychology* 52 (1953): 12.

Rosenberg, J.M. "The Cognitive Structure and Attitudinal Affect." *Journal of Abnormal Social Psychology* 53 (1956): 367-72.

Smith, M.B., Bruner, J.S. *Opinions and Personality*. New York: Wiley, 1956.

Smith, M.B. "Personal Values as Determinants of a Political Attitude." *Journal of Abnormal Social Psychology* 28 (1949): 477-86.

Tagiuri, R., Kogan, N. "Interpersonal Preference and Cognitive Organization." *Journal of Abnormal Social Psychology* 56 (1958): 113-16.

Tolman, E.C. "A Psychological Model," In *Toward a General Theory of Action*, edited by T. Parsons and E.A. Schils. Cambridge: Harvard University Press, 1951.

Woodruff, A.D. "Personal Values and the Direction of Behavior." *Scholarly Review* 50 (1942): 32-42.

Woodruff, A.D. Divesta, F.J., "The Relationship Between Values, Concepts, and Attitudes." *Educational Psychology Measurement* 8 (1948): 645-60.

Chapter 14

Appeals and Program Types

The number of people strongly attracted to any program, and "made" to tune in, depends upon three factors:

1) the number of different appeals provided by the program
2) the strength of those appeals, including appeals having greater motivating potential than others
3) the universality of appeals strongly provided, and the ability to provide strong motivation for a wide, rather than a limited, audience

In this context, reference is made to people strongly influenced by the program and wanting to tune in, rather than those people who actually do tune to the program. The number of actual listeners, or viewers, included in the audience of any program will be affected by a number of factors, including:

1) hour of broadcast
2) time of year
3) weather conditions
4) size of audience for the preceding program on the same network, or station, including cable programming
5) attractiveness of competing programs on other networks, stations, or cable channels
6) strength of appeals provided by the program

Very few programs provide all seven basic audience appeals evenly, or strongly. Most programs offer two to four, in an important or effective manner to warrant consideration. Appeals offered and degree of effectiveness will depend upon the type of program presented. Some appeals lend themselves to use in programs of certain types. The possibilities for the various appeal usage, in varying major types of programs, include the following.

MUSICAL PROGRAMS

Programs of popular music, or musical forms other than classical, provide two appeals strongly: sex appeal and emotional stimulation. Lyrics of popular music deal with love, sometimes unrequited love. Sex appeal is provided by warmth, or sexual connotation, of the voices of singers. In addition, melody and rhythm of music has relatively strong sex values. Emotional stimulation is equally important, and in some types, both are important. It is the dominant appeal in religious or patriotic music.

Equally important in popular music is the style dealing with love of home, parents, girl, or boy-friend, ranging from the "Green Green Grass of Home" to "Mary in the Morning," and from "Danny-Boy," or "The Way We Were," to "Unchained Melody," or "The Twelfth of Never." Many in the quasi-folk category deal with disasters, imprisonment, and death. "Lili Marlene," "The Last Farewell," "Today," or "The Streets of Laredo" are typical. Others deal with hopes and aspirations not yet realized: "Walk On," or the sentimental "White Christmas." Still others create a mood of reverie and thoughtfulness, such as "Tomorrow Never Comes." All old familiar songs have emotional stimulation values aside from subject matter, or musical content, by reminding listeners of past pleasant times, youth, parents, or home.

Classical music offers sex appeal values, if only in use of melody and rhythm. It also offers stronger emotional values for the refined taste in music. One secondary appeal strong in either popular or classical music is importance. Importance can be provided in the warmth of the personality of the featured vocalist or group, or those appearing on musical programs. It is stronger for the light-music type of program than in more elaborate musical productions. Comedy is occasionally used in music, such as comedy musical numbers in "nonsense" popular songs. Country/western songs and the old Gilbert and Sullivan variety provide a unique form of comedy music. In a few instances, comedy can be provided by the method of presentation. Information is extremely weak, even when an attempt is made to provide it. Information about the composer, vocalist, group or leader is sometimes given, but the effectiveness of this appeal is decidedly limited. An example of this is Dick Clark's syndicated "Rock, Roll, and Remember Program," with a series featuring a variety of musical artists on a weekly basis. One appeal never used in musical programs is conflict appeal.

VARIETY PROGRAMS

There are five major types of variety programs: comedy variety, general or vaudeville variety, amateur contest variety, country variety, and low-budget daytime or late-night variety. In each of the five, the appeals emphasized are different and varied.

Comedy Variety

The major appeal here is comedy. Sex appeal and importance are also strong appeals. Importance through name value, as well as elaborate production, can be offered. Sex appeal through use of personalities as well as use of music can be featured. In some comedy variety programs, human interest is made strong, while some form of emotional stimulation can also be introduced. In others, human interest can be weak, while emotional stimulation is nonexistent. Neither conflict nor information are present in comedy variety programs.

General or Vaudeville Variety

This type offers substantially the same appeals as comedy variety, with less emphasis on comedy itself. However, comedy, sex appeal, and importance are the strongest appeals. Combined, however, their strength is less than is possible in the comedy form, because of the lack of outstanding comedy strength. In some cases, human interest may be a relatively strong appeal, as in comedy variety forms.

Emotional stimulation is rarely a strong appeal, because its possibilities are greater in comedy forms than in vaudeville forms. As with comedy variety, neither conflict nor information can be found in general or vaudeville variety forms.

Amateur Contest Variety

This provides an entirely different set of appeals. Those strongest are human interest and conflict. As a result of the use of amateur talent, who are ordinary people, the contest elements are heightened. Comedy and sex appeal can sometimes be provided. When either is found, it is not an unusually strong appeal. Some emotional stimulation can be provided in occasional programs of this type, but its use is decidedly limited. As in the case of all variety of forms, information is never used as an appeal in amateur contest programs. But unlike the situation in cases of comedy and vaudeville variety, importance has no greater appeal value than information. Contestants are unknowns, and no elaborate production is provided.

Country Variety

This is a different arrangement of basic appeals. The strongest appeal provided is human interest. Comedy in a light form is utilized as well. Sex appeal, emotional stimulation, and importance are often provided, but these are secondary appeals. Many country music programs offer none of these appeals. A minor degree of conflict is provided, through extensive use of insult humor, and occasional use of situations where individual entertainers seem on the verge of actual battle with others. Though played for comedy values, in this case, it may be classed as conflict. As a rule, no information appeal is provided.

Low-budget Variety

This is the type normally included in daytime network schedules or late at night. The type depends upon human interest and comedy to provide the appeals. Interviews with guests and audience participants are always a part of this program form. Comedy can be provided by the master of ceremonies, and occasionally, as on the "Tonite Show," it can come from a regular personality, with material ranging from satire to slapstick. Sex appeal can be normally provided by a vocalist or musician who is part of the regular cast. Some sex appeal values can be found with the featured personality, as well. A small degree of importance is sometimes provided by use of guests. Elaborate production is completely lacking, because the low budget is obvious. Conflict is rare; however, with Johnny Carson attacking gossip magazines, there is a decided element of conflict with low-budget variety programming. Emotional stimulation and information are completely absent.

Dramatic Programs

In spite of many types of dramatic programs available, for purposes of this analysis, all can be classified under two basic headings: (1) plot drama, and (2) documentaries. All types of plot dramatizations depend primarily, and exclusively, on conflict. Conflict may dominate the entire field, including physical danger through threats to reputation, and conflict. Conflict may dominate the entire field, including physical danger through threats to reputation, and conflict of ideas. It is always present and is the most effective of all the appeals in plot dramatic programs.

However, it can be no greater than the effectiveness of conflict provided by the program. Sex appeal, human interest, and emotional stimulation follow, with human interest second. Successful plot drama deals with activities of ordinary, or real people. There is always some use of love interests, with the plot revolving around people in love, with sex appeal considered as important as conflict. If not, there will be a secondary plot involving interest of two minor characters in one another, or a hint of sex appeal provided by introduction of female characters. It might also be introduced by masculinity of the hero.

Emotional stimulation is usually provided in limited degree. In many plot dramas, use of emotional stimulation is extremely strong. With comedy-dramatic forms, comedy is the essential appeal. In this form, it may equal conflict in importance. In other programs, comedy relief will be introduced. In prestige dramatic programs, importance appeal is stressed heavily through use of name entertainers in leading parts. This is done by introduction of large casts and elaborate settings, or by dramatization of adaptations of important literary works, and use of important themes as basic plot dramatizations.

In many nonprestige plot forms, a token gesture is made for the direction of importance, by use of a name entertainer in the leading role. There are, however, many plot dramatizations where the appeal of importance does not appear in any way. Finally, in historical dramas, information may be a most significant appeal. It can also follow in true-life dramatizations or other plot dramatic programs. There is some depiction of information-appeal value in isolated adult westerns now. This comes from the fact that listeners or viewers believe productions are giving reasonably accurate accounts of life in the Old West.[1] Similarly, information may be provided in adventure drama set in foreign countries, or in dramatizations presented with medical or scientific information. Even daytime serials have been written deliberately to introduce boundaries for information appeal. In the daily daytime serial, this is one type of appeal entirely lacking.

The other major dramatic form is the documentary. Theoretically, the documentary is not considered a dramatic program at all. It lacks most elements that make plot drama attractive to audiences. In some forms of documentaries, however, there are scenes that are historical and presented dramatically. In the matter of appeals, however, the documentary, by virtue of its form, shows why ratings of documentaries are always very low. Appeals are extremely weak. The one basically strong appeal is information. All too often, however, the subject is usually on a strictly superficial level, not involving interests of the audience. Therefore, this appeal is not normally an effective one.[2]

Conflict appeal is weak or nonexistent, since there is no plot, and while conflict of ideas might be provided, the conflict is not highlighted, as is the case in a two-sided forum discussion. Human interest appeal is similarly lacking. In short dramatized sketches, there is simply no time to develop characterizations.

Characters that are presented in documentaries are not developed as people; consequently, they lack personality. Sex appeal and emotional stimulation similarly are lacking in documentaries. Comedy is never emphasized, even if it is casually introduced. Many documentaries do provide importance appeal, but in programs collectively, importance is second only to information as a key appeal. Generally, large cases are used, and in some instances, a name entertainer will act as narrator. With very few exceptions, the theme is important and vital.

Quiz, Audience, Participation, Human Interest Programs

This very general form can be broken down into two basic types: those programs containing a quiz or contest element, and those entirely lacking such an element. In the quiz or contest form, an important appeal, not necessarily the most important, is conflict. In straight question-and-answer broadcast or cable formats, conflict centering upon the battle-of-wits syndrome or the problem-solving variety, will always dominate the field. While human interest is a strong appeal, importance exemplified in tremendously large prizes, but relegated to secondary importance in elaborate production, is next to importance while human interest ranks third.

In more ordinary types of quiz programming, importance occupies a less important position. It still, however, is one of three major appeals in programs such as "The Price is Right." Strength of conflict appeal is relegated to a reduced emphasis in this programming. Aside from heavy emphasis on conflict and importance in this type of program, it conforms, generally, to other human interest or audience participation types.

Nonquiz Audience Participation

These programs depend primarily on human interest appeal. They succeed because participants are people of all races, colors and creeds. In certain types, emotional stimulation appeal is almost equal in importance to human interest, as was the case in the old "Queen for a Day" series. In others, use will be made of direct emotional stimulation, but it is secondary to human interest. Comedy is a secondary appeal in some nonquiz audience participation, particularly those of the old "Truth or Consequences" type and certainly even less in others.

In network productions, sex appeal will be introduced. Sex appeal will be provided by the personality of the master of ceremonies, which is, nevertheless, important. It can also be introduced with female assistants to the master of ceremonies. A good illustration is "The Price is Right," with vital use of very attractive models demonstrating merchandise. Frequently, there is an element of importance in these programs, through the value of the prizes given. Sometimes this element is intrinsic to the attention paid to the production. But importance is not a necessary characteristic for audience participation forms. In quiz programs, and specifically in programs such as "Queen for a Day," information appeal is introduced, with strong conflict values provided by the quizzes.

Talk Programs

For purposes of this book, attention is given only to those talk programs intended to provide information. On that basis, four types of talk programs should be considered: 1) news broadcasts, 2) informative talks, 3) persuasive talks or commentary, and 4) the panel or round-table discussion variety.

News Broadcasts

These program types offer conflict, information, and human interest as the major appeals. Conflict is essential, because news is based upon conflict situations in political squabbles, riots, threats of war, crimes, and accidents. Human interest is an element that depends on the newscaster. Human interest values are found in the news personality, in terms of his or her effectiveness. News programs provide information, because the appeal is strong, or weak, in direct proportion to the degree the listener or viewer feels involved in the events, or problems, concerning that particular piece of information.

There can be no more than slight traces of comedy, sex appeal, and emotional stimulation, although certain news stories lend themselves to and even stress these elements. The masculinity or femininity of the news personality provides a measure of the sex appeal value. Importance should be ranked after those basic appeals, while ahead of any occasional appeals.

News deals with happenings important to our society, and to the welfare of listeners or viewers. Some name value and importance is provided by important personalities in the news. Aside from name value and importance of subject matter, news programs are not suited to artificially added importance appeal. Big productions do not fit news situations.

Informative Talks

These types depend on whatever appeal can be provided by information, and omit use of any type of appeal. The result is most informative talks fail to hold listening and viewing audiences. But appeals in a well-handled, informative talk can be effective, if effort is devoted to preparation and presentation of the talk. Information that is given should deal with subjects of real and immediate concern to the listener or viewer. This does not include "odds and ends" information, but that which is vital in international, national, regional, and state news. But above all else, local variety is most preferential.

Human interest values can be deliberately introduced though references to people, anecdotes, and comments concerning problems the speaker has found. It is also found through the warmth of that speaker's personality. Conflict values can be provided, in most cases, with the speaker referring to or quoting arguments of others who disagree with his conclusions. Sex appeal, comedy, emotional stimulation, and importance normally aren't found in informational talk programs.

Persuasive Talks or Commentary

In this form, the really strong appeal comes from the information itself, since the subject matter is always a controversial public issue, and listener or viewer interest is dependent upon those conflict possibilities. Some speakers heighten conflict values by direct references to personalities through uncomplimentary references, sarcasm, and various "put-downs." Whether or not this is effective in winning audiences and bases of support depends upon target audiences and specific programs. It does heighten listener and viewer interest, however.

Next to conflict, information is the most important appeal, and human interest is relatively strong, through use of techniques mentioned previously. Other appeals are of minor significance.

Round-Tables or Forums

Everything said about persuasive talks applies equally to round-table or forum discussions. Conflict is potentially the strongest appeal, information is next, while human interest is third.

Definite efforts are made to provide human interest values, as well. Importance may be a significant appeal, as participants in the discussion may be important people, and the issues discussed could depend upon importance. Other appeals are very minor.

Not all possible types of programs have been considered; however, enough has been mentioned to suggest that for each type of program, there are two or three appeals offering opportunities. Sometimes, programs make effective use of these appeals. At other times, little or no use is made of the potential afforded by use of appeals. Any analysis of broadcast or cable programs should include a careful study of the appeals provided and the effectiveness of each of those appeals. If appeals are weak, so is the program.

NOTES

1. With any program dealing with the Old West, the producers of such a program are confronted with a form that has simply worn out. The Old West is simply stuck in the eighteenth and nineteenth centuries. In this select genre, producers must allow for: (1) a male hero, (2) a female identification character, and (3) horses or land. There is only so much you can do with such characterizations. Hollywood has tried in various ways to alter the formula, but without success.

In other respects, age groups 3–5, 5–9, 9–13, and 13–17, male and female are now interested in space series and adventure. Simply put, the space series and adventures on television and cable today utilize the basic themes of good over evil, true love, and dedication; however, there is no way you can "hitch" an X-Wing fighter outside a saloon, though Hollywood has tried! Incidentally, the bar scene in the movie *Star-Wars* is in the best tradition of the bar scene in *Shane!*

2. Much has been given to the CBS program series "*60 Minutes,*" while the program has suffered a great deal of journalistic integrity simply because the thesis is so biased. Any journalist knows that to have a formed judgment for or against someone or thing is an extremely dangerous position to take. Judgment is distorted and to "get" someone is not journalism, nor is it in line with basic reporting skills. A great deal is lost when journalistic integrity is sacrificed on the alter for ratings or "sweep week"!

Chapter 15

Strength of Appeals for Different Types of Listeners and Viewers

It is evident there is considerable variation between listeners and viewers who fall within various sex, age, educational, and cultural groups, and the degree of like or dislike for varieties of broadcast and cable programs. A program's attractiveness, for specific audiences, depends on the kind of appeals offered and the strength of those appeals. Different appeals will have varying degrees of strength for various types of listeners and viewers.[1]

Types of audiences for each specific appeal should be considered in detail. Audiences must be considered as groups rather than as individuals. Those who can be classified in one group will respond similarly to any given appeal. This doesn't mean every individual, included within any given group, will respond to a specific appeal within a program in exactly the same way as others in that group. Rather, background and past experiences will differ within groups. Listeners and viewers within a given group, however, will behave in a predictable fashion and will respond, consequently, in a presumed manner to a given appeal within a program. Allowances, however, must be made for individual variations, as they are most numerous.

THE FACTOR OF INVOLVEMENT

The degree of involvement that listeners and viewers within any particular group have depends upon the nature of the appeal, the characteristics of individuals within a group, and those experiences common to a majority of the members of that group. In the latter respect, there are substantial differences between groups or the individuals within the groups.

Those who live, or have lived, on farms have a different set of past and present experiences than those in large urban areas or in inner cities. Those who have gone to college have experiences not shared by noncollege-trained listeners and viewers. Those who attended high school have had experiences and developed interests not shared by those with more limited educations. Women, more than men, in the same age groups have a variety of different experiences and interests. Elderly individuals have experienced life in harsher times than younger people. Parents of children have sets of experiences that are not appreciated by nonparents at that equivalent age. Also, parents of children have experiences not appreciated by unmarried young adults or teen-agers. All of these various factors make for differences between groups and, based upon those experiences and present interests, may or may not have foundations in past experience.

There is a tremendous difference in degree of involvement to given appeals and difference between those in different groups, as a result of differences in past experiences or interests that will differ from group to group. This is the fundamental reason why there is variation between responses of those in one group and responses for those in another group, to a given appeal, as provided in particular programs.

There are only subjective estimates as to the responses of listeners or viewers to various types of appeals. There is no method devised for measuring the exact strength of an appeal for any given listener or viewer group. This is based upon facts found in studies of the variations of likes and dislikes of those in different groups for programs where certain appeals appear to be strong.

Conflict or Competition

This widely used type of appeal is found in a number of different forms, making any generalization with respect to it as a single appeal impossible.

Struggles Involving Danger to Life or Physical Safety

This is provided, for the most part, in thriller-type dramatizations, including westerns, crime-detective, and adventure series. Appeals are stronger for men than for women. At their peak in age groups ten to thirteen, life and death struggles will remain strong up to age twenty-five. Appeal is lower for teenagers and young adults than for adolescents. It declines steadily after the twenty-sixth year, as listener and viewer age increases; there is a good returning amount of value for men over 56.

It is stronger for relatively unsophisticated listeners or viewers. Believability decreases as sophistication increases. Consequently, it is less strong for college-educated people than those with no more than a high school education. Appeal is less strong for those in upper socioeconomic levels than those in low and lower-middle socioeconomic groups.

Physical Danger, with Strong Suspense-fear Elements

This type of conflict characterizes the "chiller" type of drama with heavy suspense values, with threats of torture and insanity, or with the presence of mental torture.

Response of women is tremendously greater with this type of appeal than that for mere physical danger. Women may respond slightly more than men, though not to an appreciable degree. The appeal has greatest strength for the same age groups as ordinary physical danger, as far as adults are concerned. The appeal strength for adolescents and teenagers is less than for ordinary physical danger. With respect to education and sophistication, the appeal is stronger for sophisticated college-trained listeners and viewers; however, the appeal values decrease as degree of listener or viewer sophistication is reduced.

Physical Conflict, without Danger to Life or Safety

This is provided in sports broadcasts. The strength of the appeal varies with all listeners and viewers, according to the degree of uncertainty with the outcome. If it is a contest so one-sided that the outcome is a foregone conclusion, the appeal is lessened. Similarly, a cable or broadcast program of an event already completed, with the outcome known, will have far less appeal value. This occurs when football, basketball, baseball, or hockey matches are time-delayed or rebroadcast over the ESPN Cable Sports Channel in other major markets after a live telecast. NCAA television and cable rules also apply here.

Appeal value varies from group to group according to the degree of involvement possible. The influence of this appeal is heightened when the listener or viewer can identify with one contestant or team. Obviously, the listener or viewer must be acquainted with and can easily identify with the sport, player, or team.

For sports generally, the appeal is decidedly stronger for men than for women, in a two to one ratio. For sports identified with colleges or universities, such as football, basketball, or baseball, the appeal will be stronger for the college-educated than for those with less formal education. Similarly, appeal for golf competition is highest for those in upper socioeconomic groups. Baseball has appeal values greater for those with high school, but not necessarily college, educations. The same is true in regard to bowling, as well as horse racing, called the "Sport of Kings".

Boxing and wrestling have greatest appeal values among those in the lowest educational and socioeconomic groups.[2] Wrestling, in particular, has greatest appeal for those lacking sophistication. Backgrounds, education, and sophistication are more important factors than listener or viewer age.[3]

Conflict appeal provided by a majority of sports will find its peak among listeners and viewers twenty-six to forty. The strength of the appeal is almost as high among younger adults, with slightly lower values among teenage boys. One must allow for differences in background, because it will decrease slowly as listener and viewer age increases, up to sixty or sixty-five years of age. In cases of wrestling, due to variations in sophistication, the appeal is greater among listeners or viewers over fifty than among those who have not yet reached that age.

Conflict Based on Threats to Reputation or Business Success

This type of conflict is second only to love conflict as the basis for plots in nonthriller and noncomedy dramatic offerings, including traditional daytime serials. Women respond more strongly than men. Reputation has greater importance for women than men, by a considerable margin. The appeal value is high in age groups nineteen through twenty-five, and twenty-six through forty, though it may be slightly stronger for the younger group. It decreases gradually among listeners or viewers forty through fifty-five, and more rapidly beyond fifty-five years of age, as listener and viewer age increases. It is a relatively weak appeal for teenagers and has virtually no appeal value whatever for adolescents and younger children. As might be expected, the appeal is strongest for those in higher educational and socioeconomic groups, with a considerable decrease as socioeconomic levels and applicable sophistication goes down.

Conflict in Love

This should be more properly classed as a form of sex appeal. This type of conflict and sex appeal show the same characteristics with respect to listener and viewer response. The appeal is greater in two-to-one and three-to-one ratios for women and girls than men and boys. It is strongest for teenage girls, particularly those over fifteen years of age, and decreases rather slowly through age groups nineteen through twenty-five and twenty-six through forty. It will more rapidly decrease thereafter, as age increases. Those with high school educations respond much more strongly than those with college educations to this appeal. The grade-school-educated group will respond much more to this appeal than those who have attended college.

Problem-solving Conflict

With few exceptions, this appeal is found in two forms: detective-mystery, or courtroom thrillers, where the viewing audience participates in attempting to identify the "culprit," or deciding what possible action a jury will take. In the second form, those programs will present problems the viewer will attempt to answer along with contestants on the program.

In the first form, the appeal value is the same for men and women. It is stronger for listeners or viewers forty-one through fifty-five, than for either younger or older adults. In the second form, the appeal value is greater for women than for men. However, the same conclusions as to age apply to both. Neither form of problem-solving conflict has any particular appeal values for young teenagers, adolescents, or younger children.

The single most important factor affecting variations in the strength of this type of appeal is the educational or intelligence level of listeners or viewers. If the problem to be solved is difficult, the appeal is weak or almost nonexistent for those in middle or lower intellectual brackets. For the appeal to have its greatest strength, the problem must be difficult enough so that with some effort, the listener or viewer can solve it and feel superior to the contestant who fails. In most "who-dun-it" mystery programs, enough clues must be provided so that anyone of ordinary intelligence can arrive at a correct decision long before the program's "hero" solves the problem. In quiz programs, questions used are difficult enough so that two out of three can be answered correctly by anyone who has completed two years of high school. As a result, the strength of this appeal, provided in most programs today, rests with those who are strictly in the middle of the educational process, with intelligence and sophistication.

In the late '50s and early '60s, a program entitled "The $64,000 Question" presented a decidedly different situation. In this type of program, questions were much more difficult and could not reasonably be answered by any ordinary listener or viewer. The listeners or viewers did not, themselves, expect or attempt to answer questions in their own minds. They merely identified with the contestant.[4] This appeal remains strongest, as a result, for the same listeners or viewers who react most strongly to appeals in regular quiz programs. Women forty-one through fifty-five registered highest, with the twenty-five through forty group close behind. Women with average educational attainments were most interested in this type of program.

Conflict of Ideas

This type of conflict is provided in discussions of controversial public issues, political campaign speeches, and political forums. The important element is listener or viewer involvement. The involvement factor depends on the degree the listener or viewer is interested in public affairs and the issues discussed. An existing interest in public affairs is found in men rather than in women. With men beyond middle age, and those with superior educational backgrounds, the appeal is strongest and highest.

The strength of the idea-conflict appeal is greater for men than women, by two to one and three to one ratios, in the fifty-six to seventy age bracket. The appeal strength is slightly lower for those over seventy and substantially lower for those forty-one through fifty-five. It will decline rapidly as listener or viewer age is decreased. It is virtually nonexistent among teenagers and extremely weak among those nineteen to twenty-five years of age. The appeal is decidedly stronger for those with less formal educations. It becomes quite low for those who have not attended high school.

These are generalizations. At varying times, an issue may be unusually high in a given community, affecting women, perhaps, more than men, or affecting younger, rather than older adults. Accordingly, in some instances, a public issue may affect those in lower socioeconomic groups much more directly than those in higher socioeconomic groups, as in the case of a strike by a labor union. In these instances, the degree of involvement strongly outweighs factors of sex, age, or education in determining the strength of the appeal.

Battle-with-one's-self conflict

This is the type of conflict provided when the hero or heroine in a dramatization is forced to make an unusually difficult decision. It sometimes involves ethics, love for family, or economic considerations. The hero, consequently, suffers considerably in deciding the course of action. This type of appeal is normally stronger for women than men. It will be stronger for those middle-aged or beyond than for younger listeners or viewers. The greatest strength lies in the age group forty-one through fifty-five, with the fifty-six through seventy group close behind. It is virtually nonexistent for teenagers or children and relatively weak for those nineteen through twenty-five.

With respect to education, cultural level, and identities, a great deal will depend upon the nature of the problem the hero must solve. In a few cases, identification with the hero is easier for those with above-average educations. This is true if the hero is an attorney, doctor, or professor. With many programs, the presentation is written so that identification is easiest for those in a middle cultural or social category. Therefore, the appeal is greatest for those with high-school training and somewhat less for those who have attended college. It is weakest for those with limited educations or in the lowest socioecomonic categories.

Comedy Appeal

As with conflict, there are numerous varieties of comedy used in broadcast and cable programming. Consequently, differences arise in the manner listeners or viewers of differing categories respond to comedy appeal. Some of the more frequently used types of comedy are listed here.

Broad "Slapstick" Comedy

In this form, everything is exaggerated far beyond limits of plausibility. The comedy characters are less than bright and, as a rule, are made to appear stupid. They are placed in ridiculous situations and do things people do not do. Therefore, elements of plausibility and believability become especially important.

Those who respond most strongly to this type of comedy appeal are children in age groups six through nine years, unless situations are beyond their experiences or capabilities. In that case, children from ten through fifteen years of age respond most strongly. Response remains consistent among teenagers as well as adults nineteen through twenty-five. From there on, it decreases for boys and girls. As age increases, difference becomes decidedly important. The response found among adults is within those categories showing the lowest levels of sophistication. The lowest socioeconomic groups and educational groups are most affected.

There is a certain degree of exception to the generalizations. If the comedian is an artist in his comedy field, his artistry may be appreciated; consequently, what he does provide is a degree of comedy appeal for more sophisticated adults, especially men.

Exaggerated Farce Situation Comedy

There is a high degree of exaggeration, though not to the extent that justifies the label of "slapstick." The leading character, however, is again less than completely bright. He makes, more or less, stupid mistakes. As a result of his mistakes, he finds himself in situations bizarre or humorous. He extricates himself from his difficulties through accident rather than use of any measure of innate intelligence. There are varying degrees of this exaggeration of character and situation used in different programs. There will always be a strain on the credulity and credibility of any reasonably educated adult.

As a result, the greatest appeal value for this type of comedy is among children ten through thirteen years of age. It is high among children six through nine years. Beyond the age of thirteen, as age increases, there is a steady decrease in appeal value. The rapidity of the decrease that takes place depends upon degree of exaggeration used in characterization or situation. The appeal is greater for those adults with lower levels of sophistication or education than those with sophistication. Among children ten through thirteen years, the appeal is equally strong for boys and girls. As age increases, the strength of the appeal becomes greater for teenage girls and women than for boys in their teens and men.

Realistic Situation Comedy

This is the type of appeal provided in situation comedies or comedy drama, where a strong effort is made to keep characters and situations believable. The leading character doesn't always act intelligently, otherwise, there would be no plot. In general, the listener or viewer feels that while people and situations are somewhat caricatured, the situation is still one that could happen. Since family situations are used, and home atmospheres are created, situations paralleling those happening in many families are utilized.

The strength of the appeal is greatest among children ten through thirteen years and for girls in particular. The appeal level remains strong and high among teenagers and adults. It decreases as listener or viewer age increases. At all ages, the appeal value is greater for women than men. The conventional use of a family situation is an important factor. The difference in appeal strength for the two sexes is a considerable one. Most situation comedy programs of this type are written deliberately for common viewers. Consequently, appeal values are highest among those who have attended high school, but not college. It is less for those with no more than a grade-school education and lowest among college-trained and more sophisticated listeners or viewers.

It is possible for the situation depicted to be one familiar to those with better educations. This might produce a change in the strength of the appeals among the college-educated and those with more limited educational attainments. In programs where this happens, there is a corresponding shift in age where the appeal is strongest. A more sophisticated type of appeal is most attractive to young adults, rather than to adolescent children.

Gag Comedy, Using Characters

Character comedy is provided by spoken lines rather than situations. A comedian, or comedians, appear as "odd" characters, such as hillbillies, foreigners, and overly loquacious women, as well as other stylized characters. Comedy is invariably broad, arising from errors in pronunciation or language usage that makes for play on words. It can come from a comedian's incorrect interpretation of the current American scene or use of unusually broad dialect, as well as from exchanges of insults.

This form of comedy has greater appeal values for men than women, in a two to one ratio. Strongest appeal values are directed to unsophisticated listeners or viewers. As a result, it is most attractive to listeners or viewers in lowest educational and cultural groups. The appeal is strongest for listeners and viewers fifty-six through seventy, and decreases in strength as listener or viewer age decreases. There is an upswing in this strength among boys from ten through thirteen years of age, or in younger teen-agers.

Gag or Line Comedy, by Straight Comedians

Bob Hope and Johnny Carson are best known illustrations for this type of comedy. It occurs in the "stand-up" sections of their programs, normally following the program opening. In the middle of Bob Hope television specials, there are gags in virtually every line, including plays on words, references to situations highly unpopular with studio audiences, mild uses of insults directed at some other performer, and similar material. Carson uses this methodology in his monologue portion of the "Tonite Show." No use of dialect, misunderstanding of statements made by others, or humor based upon errors in pronunciation or grammatical usage are made.

The appeal of straight gag comedy is definitely stronger for men than for women by a two to one ratio. However, it is different from character gag comedy in age and degree of sophistication of listeners and viewers. The strongest appeal value is usually for men in the nineteen through twenty-five age group, and strong for those for twenty-six through forty. The appeal decreases rapidly as listener or viewer age increases. It holds a moderate level for teenage boys and is lower for younger adolescents or children. The appeal has greatest value for those who have attended high school, but not college. It is second in rankings for those with some college training. It is lowest for listeners or viewers with formal educations that end before entering high school.

Sophisticated Comedy

This exists in two forms and is rarely found in broadcast programming today. First, there is the clever line variety provided by dialogue in some dramatic plays on stage, as well as in presentation specials, that are rarely provided in broadcast or cable offerings. It is an intentional and self-conscious sophistication. The same type is used in some panel programs, or by some comedians, with stress on sophistication. Audiences think, as comedians say the lines, *Wasn't that clever?* This type of sophistication has greater attraction for women than for men, although differences are negligible. It is decidedly stronger for young adults than older listeners or viewers. Comedy values are very low for listeners or viewers beyond the age of thirty to thirty-five, even to a point of negative reaction to the comedy material. Most importantly, there must be an attitude of sophistication on the part of listeners and viewers. As a general rule, the appeal is strongest for those with college training than for those less educated. Not all college-trained listeners or viewers appreciate sophistication; it is for those who consider themselves sophisticated in outlook and taste.

The second form of sophisticated comedy is quite different from the first. In fact, the label hardly fits this type of comedy. Reference is made to the comedy found in current news events, especially in political fields. A gentle irony, in most cases, is directed at the foibles of politicians or political leaders. Will Rogers and Fred Allen were outstanding exponents of this type of comedy. It is extremely rare today, if not completely missing from broadcast and cable programs, in the entire American system of broadcasting and cable.[5]

When this type of comedy is found, its appeal values are stronger for men than for women. It has a high ranking for men in middle-age groups forty-one through fifty-five, since interest in public affairs is high at this age, as well as for those listeners or viewers with college educations. But comedy appeals for this type may be frequently low for those who think of themselves as sophisticated listeners or viewers.

Sex Appeal, or Love Interest

Several types of sex appeal are found in broadcast and cable programs, requiring separate treatment.

Use of Deliberately Sexual and Physically Attractive Female Entertainers

Reference is made to the appearance of women with seductive vocal mannerisms or attractive physical characteristics that are intentionally flaunted during the time the character or characters are on camera. Sex may be emphasized in costume, mannerisms, voice quality, or in lines spoken. At first thought, it might be concluded this type of sex appeal is stronger for men than women. This is, however, open to question, as it is possible that introduction of sexual female entertainers may have a greater appeal value for women than men, on the notion that women are much more preoccupied with sex than men.

Whether the appeal is stronger for women or men, there is no attempt to reach any final conclusions in this book. Each listener and viewer must decide for himself whether appeals are greater for women or men. We can, however, generalize the appeal to be definitely stronger for listeners or viewers in their late teens and early twenties than for those in any other age group. Strength of the appeal decreases as listener or viewer age increases and is decidedly weak among listeners or viewers in the oldest age groups. In this age classification, however, it is not as weak as among children under ten years of age and boys in the ten-to-thirteen age classification.

Educational level seems to have little effect, but the appeal is slightly stronger for high-school trained listeners or viewers than for those with greater or smaller amounts of formal education. The differences with this educational level can be significant, depending upon viewing habits and educational backgrounds.

Deliberate Use of Physically Attractive Male Entertainers

This happens in broadcast and cable programs on the male side, as well as female. At one time, there was a character known as "the Continental," featured on late-night radio programs, with a sexy voice and French accent, who simply "made love" to all the women listeners throughout his entire program. Introduction of a sexually attractive male character does provide a type of sex appeal readily classified. The appeal is strongest for girls in their middle teens and remains moderately strong for women up to ages thirty through thirty-five.

It also includes those classified as "romantics," who have passed that age. Those in middle-educational groups respond more strongly than those who have attended college or those with no more than an eighth-grade education. Individuals in the latter group might not be able to believe this appeal is directed at them. Appeal for males, regardless of age and education, is almost nonexistent. This is not as open to question as female response is to female entertainers.

Sex Appeal Provided by Warm, Friendly Personalities of Entertainers

Warmth of personality by an entertainer, male or female, has a degree of sex appeal value. The degree is difficult to estimate, because this type of sex appeal merges with human interest appeal. The fact that the hero of a dramatization, master of ceremonies for a program, vocalist, or some entertainer is attractive, reasonably virile, and muscular, with a warm, pleasant voice, will give an element of sex appeal. Similarly, the fact that a female entertainer or personality is attractive in appearance and characterized by a pleasing vocal style gives her definite sex appeal values, even if there is no degree of sexuality, per se, in appearance or character.

Response to this type of sex appeal can be marked between that provided by sexually attractive entertainers and that of human interest appeal. Appeal resulting from the use of a male character is decidedly stronger for women than men. Use of a female entertainer is nearly the same, in response, for women and men. Recent research shows that it might be stronger for women than men.

For listener or viewer age, highest appeal value is in age groups older than that characterized by physical sexual appeal. It shows highest value in the group twenty-six to forty, while remaining strong for the nineteen through twenty-five age group. There is rapid decrease in appeal values for those over forty years of age. The warm personality appeal increases with less evident value for listeners or viewers in the teenage groups. It is doubtful whether education and socioeconomic factors have any particular effect on the strength of this appeal.

Sex Appeal Provided in Love Stories

This type of appeal is provided primarily in dramatic programs where two characters are in love. It can be found in certain audience-participation programs and stories told by audience participants interviewed. It may appear in other circumstances, as well. Response is much stronger for women and girls than for men and boys, by two-to-one and three-to-one ratios.

Listener or viewer age factors show a response that is strongest among girls in their middle teens and remains strong up to the age of twenty-five. It weakens for listeners or viewers twenty-six to forty and declines rapidly after the fortieth year, as listener or viewer age increases. This is the appeal provided with people in love. Most programs target the appeal for the young, late teen-agers, or those in their twenties. But if participants in the program are older, identification is facilitated for listeners or viewers closer to the age of the characters portrayed. The tendency toward finding response highest among teenagers may be offset by age requirements. At all ages, however, the basic strength of the appeal is not greatly influenced by differences in educational or cultural levels of listeners or viewers.

Those in all educational groups respond, and when ages of participants are above forty, the effect is negligible. Dramatic programs, those where sex appeal rather than conflict is the dominant appeal, seem to be best liked by women in higher education groups. There can be no generalizations as to influences of education on the strength of this type of appeal.

Use of Music with Lyrics Dealing with Love or Courtship

Love, or some aspect of love, is the theme of 95 percent of all vocal music, whether currently popular, or that music popular up to a half-century ago. Considering musical selections collectively, sex appeal values are extremely strong, and decidedly stronger for women than for men. It is at a peak among the young to the middle group of teenagers. It declines regularly in age groups beyond forty and with increases in age. It is stronger for those in middle educational categories than for those with either higher or lower educational attainments.

There are some exceptions in music. For example, sophisticated lyrics and music have stronger appeals for the college-educated than for less-educated listeners or viewers. Peak response to country music, including that dealing with the lyrics of love, will be found among those listeners or viewers in the lowest socioeconomic groups. Music taken from operas to broadway musicals tend to be best liked by listeners or viewers who have attended college. Most important of all, standards first popular with those ten through forty years of age undoubtedly have greatest participation value, and consequently, the strongest sex appeal value comes from listeners or viewers who were of high school age at the time the musical selection first attained popularity.

Being of high school age at the time the number attained popularity, and in the process of falling in love for the first two or three times, gives the music a psychological factor associated with pleasant experiences. While we can generalize about types of listeners or viewers for whom appeals are strongest, generalizations do not hold in certain types of music. All types of music with lyrics dealing with love or courtship, however, have much stronger appeal values for women than for men.

Music, as Music

Musical selections that have no lyrics, or lyrics not dealing with love, still have a degree of sex-appeal value. Music that provides strong rhythm, even the constant beat of a drum, has a varying amount of sex appeal value. The appeal is stronger for women than men, and for the young rather than old. The highest values are for girls in lower or middle teens. Education is not an important factor; however, research has shown that the lower the sophistication level of the listener or viewer, the more strongly he will respond to this particular appeal.

Human Interest

While a variety of factors make for human interest value in a program, relative strength of this appeal for different types of listeners or viewers remains much the same, regardless of the means provided. For this type of appeal, only one additional factor requires discussion.

Human Interest

For all situations, this appeal is decidedly stronger for women than for men. Appeal values will be stronger for listeners or viewers over forty years of age than for younger groups. The peak of the appeal is the fifty-six through seventy age group, with listeners or viewers forty-one through fifty-five responding as well. Those twenty-six through forty show a fair degree of response, but below that, and for those nineteen to twenty-five, there is very limited response.

Human interest is extremely weak for teenagers and practically nonexistent for adolescents and younger children. With respect to education, the appeal is weakest among those with the highest educational attainments; it is very strong for those with high school educations and slightly stronger for the less-than-high-school-educated group. Some modifications exist in certain circumstances. For example, if the entertainer providing the appeal is quite young, identification will be facilitated for younger listeners or viewers. Though the appeal values are still greatest among listeners over forty, the loss in appeal value for younger age groups becomes much less than would otherwise be the case. Similarly, if entertainers providing human interest are representative of one of the higher socioeconomic groups, identification for listeners and viewers in those equivalent groups will be made easier. Although appeal value is strongest for listeners and viewers in the lowest socioeconomic categories, there will be little decrease in strength of the appeal for listeners or viewers, as socioeconomic status of the listeners or viewers rises.

Emotional Stimulation

This type of appeal may be provided in a number of different ways. The various methods, or forms, should be given separate treatment.

Common Types of Emotional Stimulation

The most widely used methods of providing emotional stimulation values in a program are by special appeals to sympathies, use of babies, very young children, or the elderly. References to home, and all the term denotes, as well as appeals to religious interests, are commonly used in a variety of programs. All of these types of emotional stimulation are much alike in the manner listeners or viewers of these different types respond. Appeals are invariably stronger for women than men and stronger for women in lower socioeconomic categories than those in higher educational or income brackets.

They are stronger for listeners or viewers over forty years of age than for younger listeners or viewers. The greatest appeal value is found with listeners or viewers fifty-six to seventy, with strength of appeal somewhat less for those from forty-one to fifty-five years of age. There is a considerable drop in effectiveness for the appeal for listeners or viewers twenty-six to forty, and effectiveness is quite weak for those nineteen through twenty-five. It is practically nonexistent for teenagers and completely lacking for adolescents and younger children.

Special Appeals to Sympathies

If the listener or viewer is made to feel sorry for a particular participant in a program, the factor of involvement comes in strongly. Listener or viewer involvement is greater for those members of the group represented by a sympathy-arousing individual. This person may be a mother-in-law, war veteran, factory worker, farmer, a single parent, or handicapped individual, to give a few illustrations. This matter of personal involvement and identification by the listener or viewer for the sympathy-arousing individual in the program makes the strength of the emotional-stimulation appeal much greater for members of those groups represented by that individual, or those in closely related groups.

Patriotic Appeal

This is a difficult emotional stimulation type for which to formulate any general conclusions. There is evidence to believe appeal values are present for this appeal and are for entirely different groups than those responding most strongly to other forms of emotional stimulation. In normal times, appeal values follow the usual emotional stimulation patterns. In times of national emergency, however, patriotic appeal is much stronger for men than women and young adults, with men nineteen through twenty-five responding in much greater degree than older listeners or viewers of either sex. Education and socioeconomic levels are most significant factors, regionally.

Fortunately, we're not frequently in periods of national emergency. In between these crises, these special appeals for men and those younger, rather than older, listeners or viewers do not seem to be present. In ordinary times, it is difficult to suggest any particular patterns followed for variation in the strength of this appeal for different types of listeners or viewers.

Two illustrations of national emergencies in recent times brought out high concentrations of this appeal: the crises developing from this nation's involvement in Vietnam and the illegal holding of the staff of the United States Embassy in Iran. These instances brought tremendous amounts of special news coverage in broadcast and cable programming. However, the length of the Vietnam conflict, like our involvement in Korea, brought about mixed opinion and protest from the nineteen through twenty-six age groups, as well as a high concentration of discontent from black individuals in the draft-age minority group, as well.

Appeals of Nostalgia

This form of emotional stimulation calls for special treatment. Older individuals tend to recall pleasant aspects of childhood, family life, school days, and other related experiences associated with their lives in past pleasant times, while forgetting less pleasant aspects. Consequently, references in programs to those times and experiences arouse nostalgic recollections of the past, or childhood, while providing a special type of emotional stimulation.

This is far stronger for older rather than younger people. The peak level is in the seventy-year-plus group, with strength of appeal declining as listener or viewer age is reduced. It is nonexistent in very young adults, teenagers, and children. With respect to nostalgia, appeal values seem to be as great for men as for women, and differences are a result of the manner nostalgic materials are introduced and presented. The appeal varies only in minor degree among those in different socioeconomic groups.

Information

There are three types of information appeal situations in broadcast and cable programs.

Information Concerning Public Affairs, Including National News

This is a specialized type of information, widely used, identified, and classified. Included in this aspect is material provided in news broadcasts that include items dealing with national and international affairs, state or local political happenings, and discussions of important public issues. Appeal is decidedly stronger for men than women, with two-to-one ratios. It is much stronger for those in upper educational and socioeconomic groups than for those in the lower strata.

It has greatest attractiveness for men from fifty-six to seventy years of age, with appeal strength decreasing regularly as age of listeners or viewers decreases. The appeal is entirely lacking among teenage groups and nonexistent among children. There are, of course, certain types of news where appeal values are different. For the most part, however, this is considered under another heading.

Special-area Information

A second type of information relating to special fields includes everything from news concerning sports events to lectures on medical techniques. It also includes scandals involving movie actors to lessons on Japanese flower arranging, encompassing farm market reports to dissertations on development of dixieland music.

For information of this type, no generalizations are possible as to relative strength of the appeal for men as opposed to women, or older to younger age groups, and similar comparisons. Only one generalization can be made. Information in these specialized fields has moderate to strong appeal values for those interested in that subject matter.

Agricultural groups are interested in farm market reports, but only as they relate to prices of commodities the farmer or rancher has to sell or will sell in the future. Young mothers are interested in information concerning childhood development, while teenagers want characteristics of different types of top-forty rock music. Sports-minded men may be interested in sports news programs or scores of important games in regional areas. Throughout, it is difficult to make generalizations other than information appeal that is strong for members of groups collectively. Collectively, groups must be affected by the information, and where interest in subject matter already exists, the appeal will show strong results. This applies to every type of educational broadcast or cable program.

"Odds and Ends" Information

Some programs and quiz types specifically provide a variety of information on a wide range of topics not affecting the listener or viewer in any way. Sometimes, similar information is presented in forms of interesting tidbits, providing "talk" materials in disc jockey programs. Information might be of the number of square miles in Alaska, marriage rites in Honduras, or important historical events that took place on a given date.

This type of information has some appeal value; however, it is a very weak appeal. It is stronger for women than men, as well as showing strength for those with high school educations, compared to those with greater or more limited educational attainments. It will show more strength for listeners or viewers in the forty-one through fifty-five age age group, than for those of any other group. Among adults, listener or viewer age is a relatively unimportant factor.

Importance

Importance appeal is considered under three headings:

Importance of Subject Matter

The subject matter discussed may be of national or international significance, where this appeal has values for different types of listeners or viewers. The subject matter may have some strength for people affected by or interested in it, as such. Variations of importance appeal are those already outlined, but if these types of information are provided in the program, both information and importance will be provided together.

Elaboration and Costliness

Audiences are impressed by elaboration and will also respond to use of name entertainers or name personalities in broadcast or cable programs. In this type of importance appeal, the strength varies widely, according to the personalities used. Those who have strong name value in government and public affairs, such as the president, member of the national cabinet, governor, member of the United States Senate, congressman, candidate for national office, as well as heads of major corporations or nationally known labor leaders, all have name value. Importance from use of these names will depend on the possibilities of listener or viewer involvement. The same situation exists involving this appeal for any public affairs programming, despite the levels of state or national government.

Appeal strength is decidedly stronger for men than women. An even greater difference exists between response of those in top educational or socioeconomic groups and those with most limited formal education. In the matter of listener or viewer age, the appeal is greatest for older adults, specifically men fifty-six through seventy age classifications, with those aged forty-one through fifty-five coming in a very close second, and decreasing rapidly as listener or viewer age increases. This appeal is very weak for teenagers and adolescents, and virtually nonexistent for younger children.

If the name personality has a reputation earned in the entertainment world, a different situation exists. Here, the appeal is stronger for women than men. If broadway actors, popular singers, or screen actors provide the name value, greatest response will be from those with college educations, while the value will be limited for those with no more than elementary school training. For this group, highest levels of appeal are among women in upper ends of the twenty-six through forty age group, or lower ends of the forty-one through fifty-five category.

If the entertainer is a movie star, current radio or television entertainer, or popular vocalist, the appeal is still much stronger for women than men, with highest appeal centering around those with high-school educations. The age group showing greatest reaction to this type of appeal are young adults in the nineteen through twenty-five category, as well as those in lower ends of the twenty-six through forty group. An exception exists when the star is one featured in an old movie. In this situation, there is greater familiarity with the name, in this case, among those older, in the late thirties, or early forties. If the name entertainer is a currently popular vocalist, response is decidedly greater among teenagers rather than adult listeners or viewers.

The peak comes with the thirteen through fourteen age group. It remains strong through the middle and late teens and moderately strong for listeners and viewers nineteen through twenty-five. It declines rather quickly, however, after that age. Likes and dislikes for current popular music, as opposed to familiar standards, will always be established by casual relationships for listeners and viewers. High-school education is important within the age group named, while there are very few in any other educational category listed. Socioeconomic status seems to have little or no effect, while sex of the listener or viewer is extremely important. Girls, teens, and adult women respond more to this appeal, provided by current vocalists, than do men or boys.[6]

NOTES

1. Please note the previous chapter on motivations and the complete bibliography on attitudes and attitude change.
2. In the decades prior to the 1960s, wrestling was relegated to the professional variety, which took on the aspects of a three-ring circus! However, with the exacting standards set for college and university wrestling today, the sport has a high degree of sophistication and following by a large majority of university and college communities with its students, faculties, and staff.
3. See above
4. Please note the previous chapter on identification.
5. As a nephew of Will Rogers, I have seen a differentation on the term *humorist* and *comedian*. Will Rogers and Fred Allen were humorists that superseded the national scene. Will Rogers dealt with the economic, political, and social concerns of his audience through his basic preparation for broadcasts in critical periods of the history of the United States. The image in which he worked to identify himself with the concerns and image of the particular broadcast occasion and audience, is most noteworthy. In particular, his qualities and reflections were used to make a distinct brand of humor and comedy lacking today in broadcast and cable programming. Rogers and Allen shied away from the tendency to tell funny stories, preferring to relate incidents from their personal experience. Their humor and comedy came from reality, and it revealed the truth concerning the subject under discussion. As Rogers related many times:

> I use only one set method in my little gags, and that is to try and keep to the truth. Of course you can exaggerate it, but what you say must be based upon truth. And I never have found it necessary to use the words *hell*, or *damn* to get a laugh either.
>
> Personally, I don't like the jokes that get the biggest laughs, as they are generally as broad as a horse, and require no thought at all. I like the ones where, if you are with a friend, and hear it, it makes you think and you nudge your friend, and say, "He's right about that." I would rather have you do that than to have you laugh and then forget the next minute what it was you laughed at.

In his hurried day-to-day existence, Rogers' technique for preparation was to take the truth and extend it just beyond the bounds of possibility. In the context of his particular techniques for preparation, Rogers took the humorous applications and exaggerated them beyond the literal truth. Rogers' humor was found in the incongruity between the truth and Will's interpretation of it which was so illogical and inappropriate that it obviously could not be taken seriously. Although the exaggeration of the American role was unpleasant, it was not true; thus, it was a joke rich in flavor. Therefore, the humorist's reputation as a sly observer who saw through diplomatic maneuvers was further enhanced. For example, a broadcast on President Hoover revealed:

> He was chairman of the American Relief Association, and he helped feed Belgians, and a little later it was found out we was worse off than the Belgians, so they brought him home to feed us. He is always feeding somebody. Now he is feeding the Republicans. No American that ever lived can eat more than one of them can.

The story of the American Relief Association was true, but the suggestion of Mr. Hoover feeding the American people and later the Republicans was exaggerated to heighten the implicit irony. Had the humorist pointed out that the American Relief Association was feeding the Belgian people, the statement might be true, but not funny. The humor lies in the incongruity between the truth, and Rogers' interpretation of it. This incongruity, again, is heightened by the fact that the humorist sees Mr. Hoover feeding first the American people, and then feeding the Republicans. The idea that the American people are hungrier than the Belgians, and the Republicans hungrier than the American people, is enough of a distortion of the truth to throw the listener off balance, yet bring home the conviction that the American government has again made another mistake, while being bested in diplomacy.

6. Again, please note the bibliography on attitudes and attitude change.

Chapter 16

Potential Audiences and Coverage Areas

It has been emphasized that for a program to achieve the purpose it is broadcast for, it must reach a certain type of listener or viewer. They may be classed as "needed" or "useful" with respect to that purpose. The program must, also, reach a sufficient number to warrant expenditure of time and money to put that program on the air. Consequently, to the sponsor of such a program, the size of audience reached by the program and types of people making up that audience are matters of considerable importance. To deal with audiences, one must have methods of making approximations of the total size of a program audience, together with proportions of men, women and children in various sex-age groups, as well as classifications included in that audience.

CLASSIFICATIONS OF AUDIENCE

The word *audience* is not an exact word for broadcasters and cable operators. There are three types of audiences which must be considered.

Potential Audience

Any given program will be broadcast by a specific station, network, cable system, or number of different outlets in different parts of the country. The potential audience for that program includes all men, women, boys, girls, and very young children who live in homes equipped with receiving sets, in the area where the broadcast or cable program can be seen or heard satisfactorily. In other words, the potential audience includes all people living in homes where it is physically possible for the program to be received.

Two vital points should be made in this regard. First, in speaking of potential audiences, or two other audiences to be discussed, the emphasis is on people as such, not homes. Ratings are figures representing percentages of homes, while audiences are made up of people. Secondly, for potential audience, that audience will include everyone who lives in a home where it is physically possible to hear or see the program. It makes no difference as to the time of broadcast, or whether people are physically at home, awake, and with nothing to do but tune in. It makes no difference whether people are at work and not available to tune in the program.

Available Audience

The concentration for this audience is again on people, not homes. The available audience includes all men, women, boys, and girls included in the potential audience and who, at the time a given program is broadcast, are in a position to listen or see the program if they wish. Accordingly, those in the available audience must at the time of broadcast be at home, awake, and able to tune in the program. It must be physically possible for them to receive the program on the home receiving set, if they are close enough to the station or cable outlets, so that broadcasts of the program by that station or cable outlet signal are clear and effective.

This audience must also be available to tune in the program, in a place where there is a receiving set, usually at home, but in the case of radio, in a home, car, place of business, or recreational spot. In any case, radio, television, or cable, in a bar, restaurant, barbershop, beauty parlor, or any place where a receiving set is at hand and is ready to be used, can be provided to an available audience. This means the condition exists for tuning to the outlet and program possible. Obviously, every single person in the available audience must be one included in the potential audience. Otherwise, it would be impossible for them to tune in the program. Note also that the potential audience is determined primarily by set ownership and effective coverage, and will be the same, at all hours of the day. There may be variations between daytime and nighttime coverage, however, so the size of the potential audience may vary between daytime and evening. The available audience will be influenced strongly by exact times of day. The number of possible listeners or viewers, at home and awake, are substantially different at six in the morning, as compared to eight or nine on the same morning.

Actual Program Audience

The emphasis is, again, on people, not homes. The program audience is made up of men, women, boys, girls, and young children who, at the time a given program is on the air, are not only physically able to receive the outlet and available to tune in, but have actually tuned to the station or outlet carrying that program. In some cases, they are in a home or public place where the program has been tuned in by someone else. They might be listening or viewing more or less actively and attentively to that particular program.

In deciding what people should be included in the program audience, researchers must be rather generous. As listeners or viewers of the program, there are not only those who actively give attention to the program, but there are those who give only partial attention. If they are in a room where there is a set tuned to that program, even in an adjoining room paying practically no attention, they are still part of the program audience. As long as they are within listening or viewing distance of a radio, television, or cable outlet tuned to the program and receiving some of the materials of the broadcast, they are still part of the actual program audience.

It is obvious, with respect to any given program, that these three types of audiences are not similar in size. The potential audience is the largest. It includes everyone who lives in the area served effectively by the station or cable outlet carrying the program, assuming they have access to receiving sets turned on where that station or outlet can be seen or heard. The available audience, on the other hand, is considerably smaller. At three in the morning, for example, in most communities, not more than 2 or 3 percent of those in the potential audience would be awake. In the hour between 6:30 and 7:30 P.M., in the urban or metropolitan communities, there are rarely more than 70 percent of all potentials at home. Consequently, the available audience usually ranges between 2 percent and 70 percent of the total number of individuals included in the potential audience.

Not all included in the available audience will be tuning to any given program. Many are busy with household tasks, reading books, newspapers, or playing with children, while not involved in turning to programs on radio, television, or cable. With those who do, there are signals from a number of different stations or cable-access channels avilable. Many will tune to programs on other stations or access channels. Only that relatively small proportion of the available audience who are in receiving distance of the receiving sets, tuned to the particular program, are included in the program audience for that particular program.

Men, women, boys, and girls who make up the potential audience for any program must satisfy two requirements.

First, they must live in homes within the effective coverage area of the station or cable outlet carrying the program. In other words, the station carrying the program must come in, on home, automobile, or transistor receiving sets, with a good signal. This relates to the signal of the station or outlet, not to a specific program. What is meant by a good signal varies in different localities.

With AM radio stations, the signal should be strong enough when it reaches the aerial on the receiving set for the program to be heard without too much static or interference and without extraneous and irritating noise that results from having to turn the volume of the set up to too high a level. A good signal certainly implies reception without irritation from outside noises.

Regarding television or cable signals, a picture seen through a "snowstorm" is hardly considered a good signal. Distance of the home from the station or outlet's transmitter is not the only factor influencing quality of signal. A high-powered station normally can provide a good signal over an area much greater than that served by a station with low power. A less powerful signal at the antenna of the receiving set is required for good reception in country areas, rather than in cities or metropolitan areas.

With radios, there is a definite difference in the size of the area covered effectively during daytime hours and at night. These and other factors affecting station coverage will be considered at length in further passages of this book.

In network or cable programs, those in the potential audience must live in areas where some station or cable outlet carrying the program provides a satisfactory signal. Obviously, the total number of those in the potential audience will be affected by the number of stations carrying the network or cable access channel program.

Second, men, women, boys, and girls included in the potential audience must live in homes equipped with receiving sets, in working condition, and capable of receiving the signal of the station, or outlet, carrying the program. If the station or outlet carrying the program is an AM radio station, individuals in the potential audience will be equal to the total number of individuals living within the station's effective coverage area. Now, every home has one or more AM radio receiving sets. If the station broadcasting the program is an FM radio station, the potential audience will be the same, even in communities that have not had FM stations for years. The proportion of homes and cars with FM receiving sets is on the rise. On a nationwide basis, more than 80 percent of all homes and autos now have FM receivers.

A similar situation exists with respect to television. Industry estimates are that all homes in the United States now have television receiving sets. All sets are built to receive signals of television stations on the VHF channels 2 through 13, inclusive. Cable systems serve many communities, as well as providing multichannel access. Cable systems are controlled by communities of record. By that, communities contract with cable companies providing service. The contract is awarded to the particular cable facility providing superior service, as determined by the governing body of that particular community or county of record.

Nonequipped television homes, in years past, were in rural localities, seventy miles or more from the nearest city having television stations or cable outlets. In large cities, including cities having three or more stations providing television service, every home now has a television set. Every set will pick up signals of UHF stations.

EFFECTIVE COVERAGE AREAS OF RADIO AND TELEVISION STATIONS

Standards of Effective Coverage

The effective coverage area of a standard-band radio or television station is that area over which the station provides a signal strong enough to insure good reception. In case of radio, the standard engineering requirement for good reception is a signal with power of at least one-half millivolt (half of one-one-thousandth of a volt, usually written 0.5 mv) delivered at the location of the receiving set.

In case of television, a stronger signal is required. For stations on VHF channels 2 through 6, the requirement for grade B service is 0.22 mv. For stations on channels 7 through 13, it is 0.63 mv, and for UHF stations on channels 14 through 83, 1.6 mv. These are minimum standards, representing requirements for good reception in open country areas. To provide satisfactory service to receiving sets located in urban communities or small cities, the requirement for radio is 2.0 millivolts; for television, on channels 2 through 6, 2.5 mv; and on channels 7 through 13, 3.55 mv. On UHF channels 14 through 83, 5.0 mv is required. The area over which any given station provides grade B service is considerably smaller than the area that same station will provide other grades of service.

Factors determining coverage

The area over which a radio or television station provides effective coverage is determined by four major factors.

Power Used by the Station

The greater the power, the larger the area over which effective service is provided. AM radio stations are licensed to use power ranging from 100 watts to as much as 50 kilowatts or 50,000 watts. FM radio stations are permitted to use higher powers. Some operate with 300 kw power. Television stations channels 2 through 6 may use as much as 100 kw power; those on channels 7 through 13, as much as 316 kw power. UHF stations on the higher channels may use power as much as 5,000 kw, or 100 times the maximum power permitted for AM radio stations.

Frequency on Which the Station Broadcasts

This is a highly important factor in determining coverage. In general, the higher the frequency, measured in cycles per second, the smaller the area over which a station has an effective signal.

For example, even within the standard AM broadcasting band, a station using a frequency of 1500 kilocycles will cover only a third as large an area as a station with equal power but operating on a frequency of 600 kilocycles. The standard broadcast band, used by AM radio stations will cover the frequency range from 540 to 1600 kilocycles. FM broadcasting stations, on the other hand, occupy a band ranging from 88 to 108 megacycles. Since a megacycle is equal to 1000 kilocycles, this is the same as 88,000 to 108,000 kilocycles.

VHF television stations on channels 2 through 6 occupy a band immediately below that used for FM, from 54 to 88 megacycles. Those on channels 7 through 13 are in another band, above that used for FM, and extending from 174 to 216 megacycles. UHF television stations use even higher frequencies. The UHF band extends from 470 to 920 megacycles. Consequently, other factors being equal, a UHF station will cover a substantially smaller area than a VHF television station.

A television station in the channels 7 through 13 band will cover a smaller area than a television station using channels 2 through 6. An AM station covers a much larger area than either an FM station or a television station of equal power, simply because of variations in frequencies used.

Transmitter Efficiency and Height of Antenna

Elements affecting efficiency of the transmitter system are measured by qualified radio engineers.[1] One element, however, is an obvious one. The height of the antenna has a direct effect on transmitter efficiency. The relatively longer waves used in AM radio bend to follow the curvature of the earth, while the short waves used in television and FM radio broadcasting on high-frequencies travel in a straight line, so that a television station's grade A signal does not usually extend much beyond the visible horizon, as seen from the station's antenna tower. Relatively few AM stations have antenna towers more than three or four hundred feet in height. The average height of television antennas, however, is between five and seven hundred feet, measured as a distance above average level of the terrain surrounding the transmitter location. A station with an unusually high tower, or with an antenna tower located on a high hill, covers more territory than it would with a lower antenna.

Soil Conductivity and Signal Absorption

A final factor influencing the size of areas for stations providing good signals, will be the character of the terrain. In combination with the presence, or absence, of man-made structures that absorb a station's signal, a radio signal will travel a much greater distance in relatively level country than in mountainous regions. The presence of rock formations or mineral deposits below the earth's surface will absorb the signal or weaken it, so that it will travel only a short distance.

In the Plains states, particularly those from the Dakotas to Texas, the average radio station may have effective coverage over an area ten or twenty times as large as that served by stations using the same power and frequency, but located in western Pennsylvania, simply as a result of differences in soil conductivity. A similar situation exists with stations serving an area that is largely open country, as compared with one providing service to a densely populated area. Power lines, steelwork in office buildings, factories, presence of large masses of masonry, or concentrations of homes with some metal work and a considerable amount of wiring will all absorb the signal of a station.[2]

COVERAGE IN RELATION TO POWER AND FREQUENCY

It is impossible to say how large an area will be covered effectively by a given station, because location and transmitter efficiency vary too much from station to station. But the table in this section gives a good idea of the distance from a station's transmitter where good coverage can be expected, assuming ideal conditions exist with respect to soil conductivity and, in case of television stations, where use of antennas with heights of 500 feet above average terrain are common.

Figures in this table represent the number of miles from the transmitter over which a station, operating with power and frequency indicated, provides a usable signal. Daytimes for a radio station, daytimes or evenings for a television station, in open country, not in a densely populated area, and located in an area where conditions of soil conductivity are most favorable are shown above.

A grade A signal would be provided by a television station over an area with a radius of not more than one-half to two-thirds of the distance indicated. For example, a UHF television station with power of 50 kw visual, and using an antenna 500 feet above average terrain, would provide a grade A signal for a distance of only seventeen miles from the transmitter. The table makes provisions for differences only in power and frequency. Other conditions are assumed to be at least superior;

therefore, the figures given represent maximum rather than average distances.

	1000 kw	316 kw	100 kw	50 kw	5 kw	250 watts
AM radio station with frequency of 600 kc	—	—	—	305	210	95
800 kc	—	—	—	230	160	75
1100 kc	—	—	—	170	110	60
1500 kc	—	—	—	120	75	45
VHF TV station channels 7–13	—	52	46	43	—	—
UHF TV station channels 14–83	47	40	32	28	—	—

DAY AND NIGHT COVERAGES

The grade A coverage of television stations is as great during daytime hours as at night. Figures for grade B coverage, considered adequate for rural areas and small towns, as well as types with maximum figures for effective coverage, are also given in the table and are satisfactory over a much larger area at night than in the daytime.

Theoretically, AM radio station coverage should be decidedly greater at night than during the daytime. In actuality, however, for nine stations out of every thirty, the station will provide effective coverage over a much smaller area at night than during daylight hours.

Two major factors affect nighttime, as opposed to daytime effective coverage.

Skywave Signals at Night

At any given time, signals sent out by a broadcasting station will spread in every direction from the antenna. Unless directional antennas are used, signals will go north, south, east, and west, in equal degree, as well as to all points in between those major compass points. Similarly, they go down in every direction vertically, toward the ground, and straight up from the antenna again, in every direction.

Waves that go straight down, or into the ground at an angle, are quickly absorbed by the earth and have no practical value for providing a useful signal to homes located either five, ten, or one hundred miles away. During daytime hours, waves that go upward from the antenna, either straight up or at an angle, travel on indefinitely in the direction in which they start. They never reach the receivers of potential listeners. The only portion of a station's total emission of signals during daytime hours that has practical value in providing service to listeners consists of those waves going out from the station's antenna, practically parallel with the ground. During daytime hours, all reception of station signals are made possible by the ground wave (that wave paralleling the ground). The ground wave becomes weaker as it travels further from the station's transmitter; consequently, from this ground wave, different stations will provide frequency, transmitter efficiency, tower height, soil conductivity, and signal absorption. Ground wave extremity limits for stations of different types are suggested in the table.

At night, a decidedly different situation exists. Service, however, will still be provided to home, auto, and transistor receiving sets by ground waves. But in addition, part of those waves going upward at an angle from the station's antenna are reflected back toward the earth by a layer of ionized air called the "Heviside Layer." The man who first postulated the existence of such a layer of ionized air was an Englishman named Heviside. This layer forms after nightfall, at distances ranging from forty to fifty miles, and to as much as 200 miles above the surface of the earth.

Radio waves are reflected back to earth by this layer in the same way that light is reflected by a mirror. At night the effect is a skywave signal provided to areas far beyond that area relatively close to the transmitter and served by the station's ground wave. Due to variations in the height of the Heviside Layer, consistent reception every evening throughout the year is not possible and cannot be provided to all of this additional area. But at times, as a result of signals provided by skywave, a station in Texas may be heard quite clearly in Australia, or even Japan, while stations in Europe may be picked up in the United States.

Interference Problems

Unfortunately, potential coverage at night does not, for the majority of radio stations, mean better reception over a larger area than during daytime hours. In the United States, there are only 107 frequencies or channels used for AM radio, and with well over 6,000 AM radio stations assigned to those channels, or an average of thirty-five or more stations on each particular channel, there are interference problems.

Stations occupying the same frequency are separated by a small number of miles. They are far enough apart that their ground wave signals do not interfere with each other, but they are much too near one another to avoid skywave interference at night. For example, WTVN, owned by Taft Broadcasting Company, Columbus, Ohio, shares the 610-kc frequency with seventeen other stations. Fortunately, those nearest to Columbus are a 5-kw station in Philadelphia, Pennsylvania, WRCP-AM, owned by Associated Communications of Pennysylvania, and a 5-kw station in Birmingham, Alabama, WSGN-AM, owned by Harte-Hanks Communications and Southern Broadcasting Company.

Another example, WMNI-AM, owned by North American Broadcasting Company, Columbus, Ohio, is one of thirty-eight stations on the 920-kc channel. While many of these stations are off the air at night, a 5-kw station in Fairmont, West Virginia, WTCS-AM, is close enough to insure interference. WBNS-AM, Columbus, Ohio, owned by Radio Ohio, Incorporated, is one of no less than forty-four stations on the 1460-kw frequency; eighteen of that number are on the air at night, including a 5-kw station at Harrisburg, Pennsylvania, WFEC-AM, owned by Scott Broadcasting Company, which provides interference on the 1460 frequency.

WOOL-AM, in Ohio, broadcasting with 250 watts, occupies a local channel with 159 other stations, including three others in the state of Ohio, including stations at Cincinnati, Irontown, and Toledo. All but one of the 159 stations are on the air at night. WOSU-AM, WRFD-AM, and WVKO-AM, are licensed only for daytime operations, by the Federal Communications Commission.

The situation with the Columbus, Ohio, stations is typical. In every case, stations broadcasting after sunset put out a skywave signal that goes far beyond the area covered by the station's ground wave during the daytime. If there are several stations on the same frequency, the skywave signals of those other stations also come into the daytime coverage area of the station and produce enough interference so that the nighttime effective coverage area is considerably smaller than the area where an effective signal is provided during daytime hours. This applies to all AM radio stations sharing a channel with other stations at night. This group includes all stations operating with licensed power of 250 watts, 1 kilowatt, or 5 kilowatts that are local or regional stations.

There are, however, some AM radio stations protected from interference from other stations at night. These are the clear-channel stations, operating with maximum power of 50 kw. Included in this group are WKW, Cincinnati, owned by Mariner Communications Group; KDKA, owned by Westinghouse Broadcasting Company, Pittsburg, Pennsylvania; WWWE-AM, owned by Gannett Media Group, Cleveland; and WGN, owned by WGN Continental Broadcasting Company, Chicago, Illinois.

The Federal Communications Commission, the regulatory body for radio and television, has designated certain AM channels as clear channels. In each case, between one and four 50-kw stations are assigned to these channels. If more than one, distances of 1500 to 2000 miles separate the cities where 50-kw stations on the same frequencies are located.

Other stations may be licensed to operate on these clear channels during daytime hours, but these stations must leave the air at sunset. At night, only the clear-channel stations, of the 50-kw variety, operate on those particular clear channel allocations.

To illustrate, KWLG, owned by Phil Sherman, Incorporated, Wagoner, Oklahoma, must leave the air at sunset to vacate its 1520-kc frequency for the 50-kw clear-channel station, KOMA, owned by Storz Broadcasting Company, Oklahoma City, Oklahoma. Similarly, WRFD, Columbus, Ohio, must leave the air at local sunset to avoid interference with the 50-kw WCBS-AM, owned by CBS, Incorporated, New York, New York. WVKO-AM is forced to leave the air at night to avoid interference with a high-powered Canadian station in the province of Quebec. In Tulsa, Oklahoma, KGTO, owned by the Kravis Company, operating on the 1050 frequency, must leave the air to make way for a high-powered Mexican station located in Monterrey, Mexico, in order to avoid interference. Although here, this gets into agreements and accords on an international level, as well as accords that relate to regulation of the U.S. broadcast industry.

Naturally, the clear-channel 50-kw stations, between 90 and 100 of them in the U.S., have no interference difficulties at night. As a result, they get full value from the greater coverage made possible by their skywaves. 50-kw stations such as WLS-AM, owned by the American Broadcasting Company, Chicago, Illinois; WSB, Atlanta, Georgia, owned by the Cox Broadcasting Company; WWL-AM, in grand old New Orleans, owned by Loyola University; and WOAI-AM, owned by Clear Channel, Incorporated, San Antonia, Texas, cannot be heard in the Tulsa, Oklahoma area on any ordinary radio set during the daytime. However, all of these stations come into the northeastern Oklahoma area, including Tulsa, Oklahoma, with an almost local quality signal at night, as do many other 50-kw stations at even greater distances.

In summary, with respect to AM radio stations, those with power of 250, 500, 1,000, or 5,000 watts, and stations other than the 50-kw clear channel stations, have effective coverage over a much smaller area at night than during the daytime because of increased interference from other stations on the same frequency. However, 90 to 100 stations that broadcast with power of 50 kw and operate on clear channels are protected from nighttime interference problems and have benefit of the nighttime skywaves that provide effective coverage over far greater areas in the evening than during daylight hours.

This is related to AM radio stations only. Since the number of FM radio stations is growing, and perhaps as many as one-fifth of the total number that do operate use power of only ten watts, nighttime interference is not a serious problem for FM radio. Most FM stations, as a result, have somewhat larger coverage areas at night than during the daytime. The same situation exists with respect to television. On the high frequencies used for television, the bouncing of skywaves from the Heviside Layer is considerably less marked than the situation on those low frequencies used for AM radio.

Some skywave effect undoubtedly exists for television. With smaller distances covered by television stations, however, even those with unusually high power, where mileage separation between stations using the same channel is great enough, no real interference problems will exist. However, this might change with a new regulation adopted by the Federal Communications Commission, in 1982. A "drop-in" rule will add additional small stations on the same frequency as those with larger power requirements. Interference problems avoided in the past might appear as this rule takes effect in time to come.

The average television station provides grade B signal coverage over a someewhat larger area at night than during the daytime. The extent of the greater coverage is minor, when compared with greater nighttime coverage areas of 50-kw clear-channel AM radio stations.

Returning for a moment to the AM radio sphere discussed in an opening chapter, some information must be provided in regard to the Clear Channel Broadcasters' Association. Promoting radio for all America, the Clear Channel Broadcasters' Association (CCBA) has launched an intensive lobbying effort to get Congress to leave certain designated clear channels open. The CCBA claims that clear channel broadcasting is a service to the public dating back to radio's infancy and that it essentially provides more than 28 million Americans residing in the nation's vast, thinly populated regions, as well as additional millions of travelers in these areas, with their only source of radio listening after sunset.

Furthermore, the CCBA claims that these Americans who enjoy daytime broadcasting from local and regional channels would live in a radio "desert" at night, if not for the clear channel skywave signals providing wide area coverage.

The CCBA claims many Americans depend upon this service for news, weather, entertainment, sports, culture, music, education, agriculture, and religious programming. Millions who reside and travel in these areas are dependent upon clear channel stations for their only choice of nighttime AM radio service.

Over the years, the CCBA claims federal policies have permitted many of the benefits of clear channel access to be destroyed by allowing other stations to use the clear channel at night. This has brought additional radio stations and programming to larger numbers of city residents but has reduced radio programs reaching underserved rural and small town residents. The CCBA claims that preservation of the remaining clear channels is imperative as a vital link with program service, national defense, and an enhancement to the lives of millions.

Moreover, there is a claim the government's AM ceiling of 50 kw is unrealistic and must be removed to better provide reliable signals to remote areas and to assist in overcoming growing interference from higher powered foreign stations, resulting in fading, static, and increasing levels of man-made noise. The CCBA says that another alarming condition is the FCC's unrealistic power-ceiling of 50 kw, which has been in force since 1933 for class 1-A stations. Higher power (above 50 kw), according to the CCBA, will help overcome the growing interference from foreign stations and the peculiar variances of skywave service, and to counteract increasing levels of electrical noise.

Finally, the CCBA claims there are few remaining clear channel stations that provide the only nighttime AM radio service for huge rural and small town regions. More clear-channel stations should be given unduplicated status and all class 1-A stations should be allowed to use higher power in excess of 50 kw.

The Clear Channel Broadcasters' Association includes:

1. 640 Khz—KFI, Los Angeles, California
2. 650 Khz—WSM, Nashville, Tennessee
3. 720 Khz—WGN, Chicago, Illinois
4. 760 Khz—WSB, Atlanta, Georgia
5. 780 Khz—WJR, Detroit, Michigan
6. 840 Khz—WHAS, Louisville, Kentucky
7. 1040 Khz—WHO, Des Moines, Iowa
8. 820 Khz—WBAP, Fort Worth, Texas
9. 1160 Khz—KSL, Salt Lake City, Utah
10. 1189 Khz—WHAM, Rochester, New York

Address: CCB Service: 1776 K St., N.W., Washington, D.C. 20006
Phone: (202) 883-8400

NOTES

1. Licensing procedures for individuals working in broadcast or cable facilities has changed significantly over the years. Originally, a variety of tests were required for a person to obtain a license. However, the broadcast industry has been de-regulated over the years. Licensing is still required for the First Class Radiotelephone Operators' License, with Standard Broadcast endorsement.

2. Examples of this phenomena can be found in any area of the country where stations have been built prior to other structures with the aforementioned materials. In Tulsa, Oklahoma, for example, Radio Station KXXO, broadcasting on the 1300 kc AM band, has a great deal of difficulty with its signal in its home city, for the reasons listed.

WTVN-AM
42 East Gay Street
Columbus, Ohio
SMSA: Columbus, Ohio

Owner(s): Taft Broadcasting Co.
Management:
Dudley S. Taft, President
Liz Evans—Community and Pub. Aff.

Telephone: (614) 224-1271
City Population: 539,677
National Station Rep: Katz Agency
Air Time: 24 Hours
Station Type: Adult-Contemporary
Frequency and Power: KHZ 610 5,000 W
Wire Services: AP Wire, ABC Network
Network Affiliation: ABC

Bob Roof—Sales Director

WRCP-AM
2043 Locust Street
Philadelphia, PA 19103
County: Philadelphia
SMSA: Philadelphia, PA-NJ
Telephone: (215) 564-2300
City population: 1,948,600
National Station Rep: RKO Radio
Air Time: 6 AM–Sunset
Station Type: AM
Format: Oldies

Owner(s): Associated Communications of Penna, INC.
Management: Myles Berkman, Pres.
Robert Balentine—Sales D.

WSGN-AM
236 Goodwin Crest Drive, Twin Towers East
Birmingham, Alabama 35209
County: Jefferson
SMSA: Birmingham, Alabama
Telephone: (205) 942-0600
City Population: 309,910
National Station Rep: Robert E. Eastman Co.
Air Time: 24 Hours
Station Type: AM
Format: Adult Contemporary
Frequency & Power: KHZ 610, 5,000 W
Wire Services: UPI
Network Affiliation: Ind.

Owner(s): Harte-Hanks Communication Southern Broadcasting

Management: Ben K. McKinnon, Pres.
Warren Merrin—Station Mgr.
Debra Nelson—Comm. A.
Robin Carpenter—Marketing Dir.
Deborah Crumpton—Sales Mgr.

WMNI-AM
Southern Hotel, Main and High Sts.
Columbus, Ohio
County: Franklin
SMSA: Columbus, Ohio
Telephone: (614) 221-1354
City Population: 539,677
National Station Rep: McGavren-Guild Co.
Air Time: 24 hours
Station Type: AM
Format: Country/Western
Frequency and Power: KHZ 920, 100 W
Wire Services: AP

Network Affiliation: Mutual Broadcasting

Ownership: North American Broadcasting Co.
of Ownership: 100
Management: William R. Mnich, Pres.
Tim Rowe—Comm. Aff.
Libby Kirk—Continuity Dr.
Tim Rowe—Public Affairs Dr.
James Rapp—Sales Mgr.

WTCS-AM
Leonard Avenue
Fairmont, West Virginia 26554
Telephone: (304) 366-3700

WBNS-AM
62 East Broad Street
Columbus, Ohio 43215
County: Franklin
SMSA: Columbus, Ohio
Telephone: (614) 460-3850
City Population: 545,000
National Station Rep: Blair Radio
Air Time: 24 Hours
Station Type: AM
Format: MOR
Frequency and Power: KHZ 1460, 5000 W daytime, 1000 W nighttime
Wire Services: AP, UPI
Rating: 6
Network Affiliation: ABC Information Network

Owner(s): Radio Ohio, Inc.
Management: Gene D'Angelo—Pres.
Tom Stewart—Sta. Mgr.
Cindy Bertino—Continuity Dr.
Suzanne Wolery—Public Affairs Director
Thomas Stewart—Sales

WFEC-AM
900 Market
Harrisburg, PA 17101
SMSA: Dauphin
Telephone: (717) 238-5122
City Population: 68,061
National Station Rep: McGavern-Guild
Station Type: AM
Frequency & Power: KHZ 1400, 1000 W Daytime, 250 W Nighttime
Wire Services: ABC Contemporary, Mutual Broadcasting
Network Affiliation: ABC Contemporary, Mutual Broadcasting

Owner(s): Scott Broadcasting Co.
Management: Herbert Scott—Pres.
Bob O'Brian—DJ

WOSU-FM
2400 Olentangy River Road
Columbus, Ohio
County: Franklin
SMSA: Columbus, Ohio
Telephone: (614) 422-9678
City Population: 560,000
Air Time: Sunrise-Sunset
Station Type: AM
Format: News
Frequency & Power: KHZ 820, 5000 W
Wire Services: UPI, AP, AP Radio
Network Affiliation: NPR

Owner(s): Ohio State University, 190 North Oval Mall, Columbus, Ohio 43210
Bureaus: Ohio Public Radio State, News Bureau, Statehouse, Columbus, Ohio 43215
Contact: Bill Cohen, Bureau Chief

WOSU-FM
(Same as above)
National Station Rep: Don G. Davis
Air Time: 6:00 A.M.–1:00 A.M.
Station Type: FM
Format: Classical
Frequency & Power: MHZ 89.7, 50,000 W (FM)
Wire Services: UPI, AP, AP Audio
Network Affiliation: NPR

WRFD-AM
N. High at Powell Road
Columbus, Ohio
County: Franklin
SMSA: Columbus, Ohio
Telephone: (614) 885-5342
City Population: 539,677
Station Type: AM
Format: Religion/Gospel
Frequency & Power: 5000 W Daytime, KHZ 880
Wire Services: UPI, CNS
Network Affiliation: AP

Owner(s): Buckey Media, Inc.
% of Ownership: 100
Management: Dave Miller, Pres.
Michael D. Mahaffey—Station Mgr.
Christine Samms—Adv. Dr.
Myron Paul—Comm. and Public Affairs
Christine Samms—Sales Dr.

WVKO-FM
4401 Carriage Hill Lane
Columbus, Ohio
County: Franklin
SMSA: Columbus, Ohio
Telephone: (614) 451-2191
City Population: 539,677
Station Type: AM

Management: Bert Charles—Station Manager

WLW-AM
3 East Fourth, Ste. 700
Cincinnati, Ohio 45202
County: Hamilton
SMSA: Cincinnati, Ohio, Kentucky, Indiana
Telephone: (513) 251-9597
City Population: 452,524
National Station Rep: CBS Radio Spot Sales
Air Time: 24 hours
Station Type: AM
Format: Rock-Pop
Frequency & Power: KHZ 700, Clear—50,000 W
Wire Services: AP
Network Affiliation: NBC

Owner(s): Mariner Communications, 3 East 4th St., Cincinnati, Ohio 45202
Management: L. Joe Scallan, Charles K. Murdock, Presidents
Bernie Kvale—Sta. Mgr.
Marsha Edgar—Continuity Dr.
James Morris—Min. Public Affairs
Kurt Scholle—Research Dr.
Jim Meyer—Sales Dr.
Marian Shumate—Spec. Events Dir.

KDKA-AM
One Gateway Center
Pittsburgh, PA 15222
SMSA: Pittsburgh, PA
Telephone: (412) 392-2200
City Population: 550,000
National Station Rep: Radio Advertising Reps.
Air Time: 24 hours
Station Type: AM
Format: Other
Frequency & Power: KHZ 1020, 50,000 W
Wire Services: UPI, AP, Group W
Network Affiliation: NBC

Owner(s): Westinghouse Broadcasting Co., 90 Park Avenue, New York, New York 10017
% of Ownership: 100
Bureaus: Washington News Bureau
1625 K St. NW
Contact: Jerry Udwin Bureau Chief
Management: Daniel Ritchi, Pres.
B. J. Leber—Ad Mgr.
Phil Brown—Sales Mgr.

WGN-AM
2501 Bradley Place
Pittsburgh, PA 15222
SMSA: Pittsburgh, PA
Telephone: (312) 528-2311
City Population: 336,695
National Station Rep: WGN Cont'l. Sales Co., Christal Company
Air Time: 24 hours
Station Type: AM
Format: Personality/talk/sports
Frequency & Power: KHZ 720, 50,000 W.
Wire Services: AP, UPI, CNB
Network Affiliation: Ind.

Owner(s): WGN Continental Broadcasting Co.
Management: D. T. Pecaro, Pres.
Wayne Vriesman—Sta. Mgr.
Kenton Morris—Ad. Mgr.
Dick Jones—Comm.

KOMA-AM
P.O. Box 1520
Oklahoma City, Oklahoma 73101
SMSA: Oklahoma City, Oklahoma
Telephone: (405) 794-1573
City Population: 366,481
National Station Rep: John Blair & Co.
Air Time: 25 hours
Station Type: AM
Format: Country-Western
Frequency and Power: KHZ 1520, 50,000 W
Wire Services: UPI
Network Affiliation: Ind.

Owner(s): Storz Broadcasting Co., Kiewit Plaza, Omaha, Nebraska 68131
% of Ownership: 100
Management: Robert H. Storz, Pres.
Woody Woodward—Sta. Mgr.
Pamela Fox—Comm. Aff.
Jan Taylor—Continuity
Don Wilbanks—Research
John Rogers—Sales Mgr.

WCBS-AM
51 W. 52nd St.
New York, New York 10019
County: New York
SMSA: New York, New Jersey
Telephone: (212) 975–4321
City Population: 786,776
National Station Rep: CBS Radio Spot Sales
Air Time: 24 hours
Format: News
Frequency and Power: KHZ 880, 50,000 W
Wire Services: AP, UPI, Reuters
Network Affiliation: CBS

Owner(s): CBS, Inc., 51 W. 52nd St., New York, New York
Management: Robert L. Hosking, Pres.
James McQuade—Sta. Mgr.
Teresa Ditmore—Comm. Aff.
Donald O. Gorshi—Sales

WLS-AM
360 No. Michigan Ave.
Chicago, Illinois 60601
County: Cook
SMSA: Chicago, Illinois
Telephone: (312) 984–0890
City Population: 336,695
National Station Rep: John Blair Co.
Air Time: 24 hours
Station Type: AM
Format: Top-Forty
Frequency & Power: KHZ 890, 50,000 W
Wire Services: AP, UPI, City News Bureau, U.S. WEA
Network Affiliation: ABC

Owner(s): American Broadcasting Co., 1330 Ave. of the Americas
Management: Elton Rule, President
Leta Wilson—Continuity
Regina Hayes—Pub. Aff.
Karen Cavaliero—Res.
Craig B. McKee—Sales
John Cravens—Sales

WSB-FM
1601 Peachtree St. NE
P.O. Box 4146
Atlanta, Georgia
County: Fulton
SMSA: Atlanta, Ga.
Telephone: (404) 897-7500
City Population: 496,973
National Station Rep: The Henry Christal Co.
Air Time: 24 hours
Station Type: AM
Frequency & Power: MHZ 98.5, 100,000 W. (FM)
Wire Services: CBC, AP, UPI
Network Affiliation: NBC

Owner(s): Cox Broadcasting Group
Management: Clifford Kirtland, Pres.
Jack Lenz—Sales Mgr.
Ann Cooper—Sales Mgr.

60 WWL-AM & FM
    1024 N. Rampart St.
    New Orleans, LA.
    SMSA: New Orleans, La
    Telephone: (504) 529-4444
    National Station Rep: The Katz Agency
    Air Time: 24 hours
    Station Type: AM
    Frequency & Power: MHZ 101.9, 50,000 W
    Wire Services: AP, UPI, Wall St. Journal Wire Serv.
    Network Affiliation: CBS

    Owner(s): Loyola University
    Management: Rev. James Carter, Pres.
    Raymond M. Curo—Sales

61

62 WOAI-AM
    1031 Navarro St.
    San Antonio, Texas 78205
    County: Bexar
    SMSA: San Antonio, Texas
    Telephone: (512) 226-9331
    City Population: 654,153
    National Station Rep: Buckley Radio Sales, Eastman
    Air Time: 24 hours
    Station Type: AM
    Frequency & Power: KHZ 1200, 50,000 W. AM
    Wire Services: NBC, WOAI Local Bureau, CBS
    Network Affiliation: NBC, CBS

    Owner(s):
    Clear Channel Communications, Inc.
    Management: L. Lowry Mays, Pres.
    Rex Tackett—Sales

Chapter 17

Available Audiences and Selection of Broadcast Programs

POTENTIAL AND AVAILABLE AUDIENCES

The potential audience of a radio, television, or cable entity includes all individuals living in homes equipped with receiving sets, when those homes are located within the effective coverage area of the station or linked to a cable outlet. The potential audience includes 100 percent of the men, women, boys, and girls living in those homes. It does not vary at different times of the day. On the other hand, at different times, not all of those people who make up the potential audience will be available as listeners. At some hours, most of those included in the potential audience will be asleep. During most of the day, employed men and women will be out of the home and at work. On weekdays during the school year, children of school age will be in school. Even those persons at home and awake at any given time may be working in the garden or outside chatting with a neighbor. They are not available to listen to radio, or watch television or cable programs.

The available audience of a radio, television, or cable outlet includes all men, women, boys, and girls who are included in the potential audience, and who are, also at a given time a program is on the air, are at home, awake, in the house, and available to listen or watch, if they choose to do so. The size and composition of an available audience varies at different hours during the day. Accordingly, enough people are always out of the home and, at night, are asleep, so the available audience, even at its peak, includes no more than 75 to 80 percent of individuals who make up the potential audience. And during much of the time during the day, it includes less than 40 percent of those individuals.

HOUR BY HOUR CHANGES IN AVAILABILITY

During daytime hours, children make up substantial portions of the available audience. Results of this fact show up strikingly well in large audience ratings and percentages of total homes attracted by television or cable programs aimed, principally, at young children. Variations in numbers of preschool children available during the daytime indicate that in many households, the early afternoon nap requirement is enforced. In pleasant weather, children spend part of their time out of the house, in outdoor play and recreation, which accounts for changes in availability.

The number of teenaged boys and girls available in the fourteen- to eighteen-year-old-bracket suggests that managers of some radio stations who aim programs chiefly at teenagers may not realize how few numbers of boys and girls in the teenage category make up the total population. At every single hour of the day, more women are available than men. There are reasons for this. First, the number of women in the total population of American cities is nearly one-sixth greater than the number of men. Also, there are more men than women living on farms, and the numbers are not that different in small towns. However, women, more than men, leave rural areas to find work in cities, because there is little opportunity for employment for women in farm communities or in small towns. In addition, women, on the average, live five to ten years longer than men.

A second factor is housewives are still tied to the home, more than men, by the presence of children. The greater freedom women enjoy today has not altered basic roles that significantly. Men, more than women, have duties that keep them out of the home during evening hours, as well as during the daytime. At every hour of the day or night, up to midnight, in cities, more women are available as radio, television, or cable listeners and viewers than men.

THE MOST VALUABLE AUDIENCE

The purpose of most broadcast and cable programs is to sell goods and services to consumers. From that point of view, the most valuable audience for broadcast or cable programming consists of women in the twenty-six to forty age group, while the second most valuable audience includes women aged forty-one through fifty-five. Women are valuable because they do a majority of marketing for the family, especially goods generally advertised on radio, television, or cable. The two age groups are essential because those families where the housewife is between twenty-six and fifty-five do 75 percent of the total buying of consumer goods.

Households where the housewife is between twenty-six and forty years of age are often those where children under ten years of age are present, as well as 50 percent of the children and teenagers from eleven to eighteen years of age. Getting a new home started will involve women. In addition to raising a family, with or without children, women will make crucial decisions in the purchase of products from furniture to breakfast cereal and clothing to cleaning preparations.

The second age group of housewives will include those from forty-one through fifty-five. These women preside over families that are smaller but still have one or more children of junior-high or high-school age. Families in this group have higher average annual incomes than any other group. In addition to buying necessities that characterize both family age groups, families in the older group are the best market for quality merchandise and items in the luxury category.

For these circumstances, a good deal of attention should be given in considering availability of listeners or viewers to times during the day when women twenty-six through forty and those forty-one through fifty-five are at home and free to listen, or see, broadcast or cable program offerings.

SELECTION OF BROADCAST PROGRAMS

When it is important to reach a particular type of listener or viewer, to achieve the purpose of a radio, television, or cable program, the method used to select programs and the member of the family group who makes that program selection for available members of the family are matters of basic importance.

HOW PROGRAMS ARE SELECTED

A first consideration is the method whereby a given program is selected to be seen or heard in any particular home. In general, methods of selection may be classified under three headings.

Carry-over Selection

The car radio or home television set is already tuned to some station or cable channel in this process. As a result of the attractiveness of a preceding program broadcast by the outlet, there is no particular reason to make a change; therefore, the set is allowed to remain tuned to that station or outlet. The family group available in the home or auto will listen or watch whatever that station or outlet offers. However, if a program is completely unsatisfactory, the carry-over process is no longer effective. Either the set will be turned off or returned to a different outlet providing a more satisfactory program.

Dial-twisting Selection

In this situation, no member of the family or car group is especially interested in any program known to be scheduled at a given hour. Either the family set has not been turned on prior to the start of the period considered, or it has been tuned to a station or outlet now carrying a program disliked enough to require search for a different and more satisfying program.

The set is tuned to various frequencies or channels. The program carried on each outlet on the dial or cable switcher is tried out briefly, and a selection is made on the basis of the program that seems most attractive to the person controlling the dial at that moment. Sometimes a variant of this process is used. The person who selects the program checks the program log of a newspaper to see what is available. Again, selection of the program is made on the basis of which program broadcast is most attractive. This is without the program selector having any strong preferences in advance for any particular program offered by the stations or outlets serving that area.

Intentional Selection

With this method, some member of the family group knows an especially well-liked program is to be broadcast at a certain hour and makes a point to remember when that program is to be broadcast. That person will tune in the station or cable outlet carrying the program at the proper time. It is possible that the set may already be pretuned to the desired station or cable access channel. Selection is still classed as intentional, rather than carry-over, if the program selected is one known about in advance.

A good part of the selection process for programs is the carry-over variety. This is particularly true in case of radio, where family home sets, car radio, or transistor may be tuned to a favorite station for two or three hours at at time. Carry-over selection is also rather common with respect to television or cable programs during the daytime. It is doubtful whether as much as 50 percent of all daytime television program selection is made on a carry-over basis. Available women and children have favorite daytime programs that are tuned in to every day. Therefore a segment of the family group is present at home to watch as part of a family group routine.

Dial-twisting is the rule if a listener is not acquainted with the radio stations available in the community. There is a considerable amount of dial-twisting selection for television or cable programs. This is true at those times of the year when regular programs, on during winter months, are replaced by summer broadcasts or cable reruns. This occurs at the beginning of the autumn season, as well. There is a certain amount of dial twisting, too, for evening hours during regular television and cable seasons. A listener or viewer may be available and desire to tune in at a time when he usually does not use the television set. The unfamiliar person will select the best program available, either by sampling program offerings of various stations or by making a choice from newspaper logs.

Intentional selection accounts for only a small proportion of the total selection of radio programs. It is less than 20 percent, and no more than 30 to 40 percent, of daytime television selection. The same daily programs are available five days a week; consequently, it is easier to be acquainted with several of the programs offered. At night, 20 to 40 percent of all television or cable programs tuned in, for the average household, are chosen on an intentional basis. For example, some members of the family know a particular entertainment feature will be broadcast at a particular time and like the program well enough to tune in at the proper hour.

The Program Selector

At any given time, the actual job of selecting the program to be tuned in on the family set, or sets, especially when either dial twisting or intentional selection is used, will be taken over by one individual member of the available family group. Within the family group that includes adult men and women, possibly teenagers or younger children, the selector will not always select the program which would be chosen by other members of the family group that are at home and available. Other family members will give a greater, or lesser, amount of attention to the program selected, based upon their individual preferences.

The program selector makes a difference in the determination of what program will be tuned in. Therefore, there is a role-playing scenario as to which member of the family group, at any given time, takes the role of selector. Accordingly, there is a problem of reaching the right audience, in relation to those causal relationships developing from the selector of those particular programs. In a large majority of cases, the purpose of a program can be achieved only if the program reaches a substantial number of a particular kind of listeners or viewers, from the standpoint of sex, age, and other qualities.

How can the broadcaster or cable operator have any hope of reaching the desired type of listeners or viewers when the program selector in one household may be a man, in another a woman, and in a third, a small child, at the time the program is put on the air?

For one possible answer, when "60 Minutes" is broadcast on Sunday evenings, the proportion of men who are allowed to select the program can be above or below average. When the Dick Clark Show is broadcast on Saturdays, an astonishingly high or low proportion of children role play as selector for the program. Therefore, broadcasters and cable operators attempt to provide a program so unusually attractive to particular kinds of listeners or viewers needed that those listeners or viewers will make a stronger-than-ordinary effort to serve as selectors of that program in order to choose the program in question.

The second possibility is to provide a combination of appeals in the program that will make it attractive. It can be unusually attractive to every type of listener and viewer; consequently, no matter who selects the program at a given hour, that person is likely to choose the program considered. Unfortunately, that does not happen too frequently either. In most cases, a program appeals strongly to one special group but not to every type of listener or viewer equally.

Chapter 18

Audiences of Specific Programs

Importance of audience information cannot be overemphasized, because a radio, television, or cable program is broadcast to achieve some predetermined purpose. To that end, it needs to reach some particular type of listeners or viewers, in large enough numbers to justify time and money spent to put that program on the air. Those who present programs, consequently, have reason to be interested in the size of audience reached and the specific makeup of that audience. This chapter deals with those two aspects of a specific program audience.

There is no exact method of determining what kind of people make up the audience of any selected program or how many people of one specific type will tune to that program. In large metropolitan areas, we have local ratings available to give approximate percentages of homes in carefully designed rating areas where receiving sets are tuned to specific programs. If a program is commercially sponsored, there are rating services that provide information as to the number of men, women, and children tuning to the program per 100 homes where the program was turned on.

Beyond this, there is no information, even in those major metropolitan markets or population centers. In making estimates of the size or composition of the program audience in a small community, where no rating service is available, or determining exact proportions in each of the several sex-age groups, the best that can be done is to make approximations based upon what is known about size and characteristics of the potential audience for the station over which that program is broadcast. There are methods by which audience size and composition for any given program may be approximated. However, any approximations made concerning size or nature of the audience for any program are based upon guesswork. They cannot be considered totally accurate in any way.

LOCAL RATINGS FOR THE PROGRAM

Estimates must start with ratings received by the particular program, in a local community where a station carrying that program is located. If the city is one where one of the recogized radio or television research companies provides a regular monthly rating for programs, based upon listening or viewing in that local community, and if this rating information is available, it can be used.

If local rating information is not available in small markets or for sustaining programs broadcast over stations or cable outlets in larger markets, the first step in making an approximation of size and composition of the program audience is to make an estimate of the rating that program might be expected to have in the community where it is broadcast.

LOCAL SETS-IN-USE FIGURES

To estimate ratings, one should begin with total listening or viewing at the hour the program is broadcast. The sets-in-use figure is the proportion of homes in the community where some set is tuned to a radio, television, or cable program. The figures in the table offer a starting point.

Hour Starting	Sets-in-use TV	Radio	Hour Starting	Sets-in-use TV	Radio	Hour Starting	Sets-in-use TV	Radio
6:00 A.M.	0.4	5.5	12:00 noon	21.00	11.8	6:00 P.M.	34.7	9.0
7:00 A.M.	3.9	12.6	1:00 P.M.	19.00	11.6	7:00 P.M.	45.7	7.0
8:00 A.M.	9.0	15.1	2:00 P.M.	17.4	9.9	8:00 P.M.	56.5	5.3
9:00 A.M.	10.9	14.1	3:00 P.M.	18.7	9.5	9:00 P.M.	60.5	4.8
10:00 A.M.	13.3	13.8	4:00 P.M.	22.3	8.6	10:00 P.M.	53.9	4.5
11:00 A.M.	17.9	12.0	5:00 P.M.	28.0	9.2	11:00 P.M.	33.8	4.1

The table shows sets-in-use figures for television and radio, based only on homes having television and/or radio receiving sets, as computed by averaging the figures for January, April, July, and October, by the author, in Tulsa, Oklahoma. The radio sets-in-use figures are considerably lower than those reported by other rating agencies. But since the author's figures are taken from data secured by use of a college journalism-broadcasting research project for television and radio receiving sets in a carefully selected sample of homes in Tulsa, Oklahoma, it can be assumed that the figures are more accurate than those provided by other means! The figures also represent Tulsa, Oklahoma, averages on a year-round basis. The actual proportion of homes using television or radio sets may be higher or lower in different communities at different times of the year.

For instance, the television sets-in-use figures for the hour 2:00 to 3:00 P.M. on weekdays is 21.0 in December but only 13.9 in September, compared with a year-round average of 17.4 given in the table. There is less month-to-month variation with respect to radio, however. But both radio listening and television viewing varies from community to community, and the sets-in-use figure is not likely to be more than one-tenth higher or lower than the national average.

These are factors, however, that may affect the total amount of listening or viewing and should be taken into consideration. There will also be peculiar listening or viewing situations that exist within any given community. This type of listening or viewing is done despite whether or not the weather is pleasant or disagreeable on a given day. Attractiveness of programs offered on a given day of the week, as compared with other weekdays, should be considered because of sophistication levels of audiences in various markets.

Figures in the table report the proportion of homes listening or viewing at various hours on weekdays, not on Saturdays or Sundays, during hours before six in the evening. Saturday or Sunday audience availability, and consequently, the listening or viewing situation, may be quite different from the situation existing during daytime hours on weekdays, Monday through Friday.

ESTIMATION OF PROGRAM RATINGS

If there is an approximate sets-in-use figure for the hour the program is broadcast, in a community where the station carrying the program is located, an approximation can be made for the rating that program might receive. Several factors, however, must be taken into consideration.

Number of Stations

In ratings study, the number of stations—radio, television, or cable access channels—located in and serving the community determines the average ratings. The average ratings of programs carried on a few stations during that hour will be 5.0, while if there are ten stations, the average rating of programs would be only 2.0. If the program is broadcast on television, where there may be only one or two stations in the market, the rating for a given network program will be decidedly higher in that community than in a city where there are three or four broadcast television stations and a host of cable outlet access channels.

Prestige and Popularity of the Station

A second most important factor is the extent that people living in that community or regional area will habitually listen to or view a station carrying the program, as compared with the amount of listening or viewing they will give to other stations. The table in this section illustrates this factor by showing average ratings of programs carried on each of the commercial television stations in Tulsa, Oklahoma, on weekdays during the months of January and February, 1982, as reported by the author's researchers.

Average Rating of Programs On:	Daytime	Evening	Morning 6–8	8–10	10–12	Afternoon 12–2	2–4	4–6
KTUL (ABC)	9.3	22.7	3.9	7.5	11.1	12.5	7.2	13.6
KJRH (NBC)	10.2	15.9	3.0	9.8	10.9	9.4	10.2	18.3
KOTV (CBS)	4.0	9.2	—	5.2	2.3	3.3	4.8	4.3
KOED (PBS)	2.0	1.5	—	—	1.0	.5	.5	2.0
KOKI (UHF) Ind.	2.5	1.5	.5	1.0	1.5	.5	.1	1.5
KGCT (UHF) Ind.	.5	1.0	—	—	.5	.1	—	.1

Average Rating of Programs On:	Evening	Evening 6–8	8–10	10–12
KTUL (ABC)	22.7	22.5	31.1	14.6
KJRH (NBC)	15.9	20.5	17.6	9.6
KOTV (CBS)	9.2	13.5	11.2	3.0
KOED (PBS)	1.5	6.5	5.1	1.5
KOKI (UHF) Ind.	1.5	2.1	3.5	3.0
KGCT (UHF) Ind.	1.0	.5	1.0	.5

As noted above, figures are from the local Tulsa area, reported by the author, and are nothing more than a college survey and study conducted in a rather haphazard fashion. The major purpose of the table, however, is to illustrate the fact that in any given community, some stations habitually pull larger audiences than others. In this case, KTUL, for years locally owned in northeastern Oklahoma by a most celebrated broadcaster, James C. Leake, Muskogee, Oklahoma, has had a great deal of popularity.

Allowances must be made for the relative attractiveness of that station carrying the program in estimating the local rating for that program.

Strength of Appeals Provided by the Program

Another factor that must certainly be taken into consideration concerns appeals provided by a program. A program offering usually strong appeals will, sooner or later, attract a large audience and consequently have a relatively high rating, regardless of the general popularity of the station carrying the program. Station popularity works in a most peculiar fashion. Assuming two programs with equally strong appeals are broadcast over competing stations, the program carried by the station with greater overall popularity will have a higher rating than will the other program. But aside from station popularity, the strength of the appeals in that specific program is vitally important.

Additional Factors

Other criteria must also be taken into consideration, including size of the audience attracted by the immediately preceding program on that station. This will, consequently, create an opportunity for carrying the audience over. General attractiveness of competing programs, broadcast at the same hour on other stations, must also be considered. The question of which member of the family group acting as program selector in most homes at the time the program is broadcast, and of whether or not that program offers special appeals for the type of individual serving as selector, must be considered. From these factors, it is possible to make an estimate of the rating for a given program under consideration and of what it would receive in the community where the station carrying that program is located.

Homes Reached by the Program

A rating is a figure representing the percentage of homes that have sets tuned to the program rated during the average minute the program is on the air. Having a rating figure, either one provided by a commercial rating service or an estimated rating, will prepare us for the next step, which is to convert that rating into a total number of homes where people are listening or viewing the program. Here, it is necessary to take into account the size of the area covered by the station carrying the program. The total program audience, whether in terms of homes or individual listeners or viewers, should therefore be broken into two elements.

First, there is that part of the total audience located in the home community where that station is located. Secondly, there is that portion of the audience located outside the home community. The outside portion of the audience, located away from the home community or primary coverage area, is called the outside audience, or area of secondary coverage and influence.

Home Community Audiences

A local rating for a program is based upon the local community, which is the area of dominant influence where that station or outlet carrying the program is located. If you have a local rating for the program, with figures showing the number of homes in the community equipped with radio and/or television sets, the number of homes in the home community audience can be determined by multiplying the total number of radio- or television-equipped homes in the community by the rating figure, expressed as a percentage.

For example, in the metropolitan Tulsa, Oklahoma area, there is a total of 200,000 homes equipped to receive both radio and television programs. A program with a rating of 10.0 would be heard in 10 percent of 200,000, or 20,000 homes equipped to receive that program in the metropolitan Tulsa area. Accordingly, if the rating for a radio program in the Tulsa, Oklahoma area is 3.5, the program will be heard in 3 1/2 percent of 200,000 homes, or 7,000 homes in the metropolitan Tulsa area.

The Outside Audience

Estimating the area of dominant influence or home community audience with the number of homes tuned to the program is relatively easy, if a reasonably accurate rating figure for the local area is available. However, estimating the number of homes in the outside audience for the program is difficult. For that outside area, there is no rating figure for the program. One does not know how many outside homes the station is able to provide, and there is little or nothing known about the competitive situation from other stations. The strength of appeals for the program is not the same for an essentially rural audience as for an audience in the home community.

An approximation can be made, however, from the outside audience by taking into account the coverage the station carrying the program maintains, in view of its power and frequency. Secondly, the extent to which the station generally aims its programs to a rural outside audience, rather than listeners or viewers who live in the home community, should be the counterpart for the approximation.

For example, a 50-kw AM radio station on a low or medium frequency may well regularly reach two, three, or four times as many outside homes as it reaches in its home community. The approximation can be further verified if its programs are directed primarily toward a rural, rather than city, audience. The same might be true of a 5-kw station on a low frequency, located in a smaller city, if the station is programmed primarily for rural listeners. It is doubtful whether the number of outside homes reached by the average program on this station will be more than 50 to 70 percent the number of homes reached in that smaller city. The number of outside homes reached by a 250-watt station, in a rural community, will rarely be more than 20 to 40 percent the number of homes reached by the same program carried on that station in the home community. Outside coverage, by any AM station other than a clear channel station, will be decidedly smaller at night than during daytime hours. With respect to the average VHF television station, of the type located in Tulsa, Oklahoma, it would be a generous estimate to assume the outside audience, daytime or at night, will be any more than 15 to 20 percent the number of homes reached in the metropolitan Tulsa community. The area covered is small, and population is only a fraction of the population living in the metropolitan area itself. An estimate of the outside audience, therefore, is simple guesswork. There simply is no accurate information on which an estimate can be based.

Number of Program Listeners

Assuming one can be reasonably certain of the total number of home community homes tuned to the program, and having made a guess to the number of additional homes listening to the program in the outside area, the estimate of the number of individuals, men, women, boys, and girls, listening to the program is nothing more than an estimate. We do have some figures showing the average number of individuals per 100 Tulsa homes, listening or watching the average radio or television program at different hours of the day. These figures are provided in the table.

They are taken from Tulsa Junior College's Journalism-Broadcasting Department's classes in basic audience composition research, February, 1982.

Viewers and Listeners per 100 Homes Tuned to Average Programs on Weekdays During the Hour Beginning At:

	6	7	8	9	10	11	12	1	2	3	4	5	6	7	8
Programs on TV															
Total	—	162	160	142	139	144	152	150	146	155	166	167	198	210	235
Men	—	48	28	14	13	15	16	15	14	14	12	21	54	66	81
Women	—	73	57	80	86	86	89	90	90	80	71	40	62	73	81
Teenagers	—	7	8	7	7	7	7	7	6	9	24	19	15	17	17
Children	—	34	67	41	39	36	40	38	36	51	59	87	67	54	48
Programs on Radio															
Total	150	181	155	125	128	143	136	129	127	138	148	156	165	156	152
Men	75	81	55	29	28	29	33	27	28	32	43	53	69	67	65
Women	71	83	83	84	89	101	87	89	84	83	75	73	70	67	65
Teenagers	3	11	10	5	4	5	9	6	8	17	25	24	21	18	18
Children	1	6	7	7	7	8	7	7	7	6	6	6	5	4	4

	9	10	11
Programs on TV			
Total	255	188	166
Men	84	83	79
Women	92	84	83
Teenagers	16	11	3
Children	33	10	1
Programs on Radio			
Total	149	141	131
Men	65	64	63
Women	63	62	60
Teenagers	18	14	8
Children	3	1	—

Note that the number of individual listeners and viewers per 100 homes is different for radio than for television. This is significant from the standpoint of the number of children listening. The housewife giving information to our interviewers felt that children who were in the home simply weren't listening to a radio program that was turned on. Secondly, the proportion of women of child-bearing years listening to radio was low. It is evident that, at most hours of the day, there will not be more than 1.5 listeners per listening home for radio.

The conspicuous exception is in the period from 7:00 to 8:00 A.M.

In the case of television, the number of individual viewers, per viewing home, ranges from 1.4 to 1.7 during daytime hours and goes as high as 2.3 at the peak hour of availability, at night. The figures, of course, are for weekdays, and there was no information provided concerning daytime listening on Saturdays or Sundays. For an approximation of the total number of listeners who tune to a given program, the total number of homes reached by the program may be multiplied by the number of listeners per listening home, or 1/100 of the figure given in the preceding table for the hour of broadcast for that program. It must be emphasized this is nothing more than an approximation. It is based upon a guess as to the total number of homes reached in the home community and in the outside area. Similarly, only a figure representing the number of listeners per 100 listening homes tuned in to all programs, of whatever type that are available at the hour of broadcast, can be given.

THE COMPOSITION OF THE AUDIENCE

Up to this point, all of the research computations suggested have been for the purpose of estimating size of audiences listening to a given program. The number of individuals listening to that program in a given community is the solution. But if the program is to be effective in this solution stage, it must reach, in particular, the kind of listeners needed for that purpose. Programs, however, will vary in kinds of listeners they attract.

The preceding table gave average figures for the number of men, women, teenagers, and children listening to radio or watching television programs at a given hour, per 100 homes tuned to radio or television. But specific program audiences may be quite different in makeup from the overall audiences. Our research figures used were computed to give information concerning the makeup of the audience for each separate television program in the Tulsa, Oklahoma, market, and the makeup of the AM radio audience listening to two Tulsa, Oklahoma, area radio stations, during each hour of broadcast. Some of the audience-makeup figures are compared with average figures for the time of broadcast, in the table below.

LISTENERS/VIEWERS PER 100 LISTENING/VIEWING HOMES

	Total	Men	Women	Teenagers	Children
5:00 P.M. Average: Radio					
Basic Average	156	53	73	24	6
on KVOO at 5:00	158	60	79	14	5
on KRMG at 5:00	158	46	67	39	6
7:00 P.M. Average: Radio					
Basic Average	156	67	67	18	4
on KVOO at 7:00	156	75	72	6	3
on KRMG at 7:00	161	60	58	38	5
4:00 P.M. Average: Television					
Basic Average	167	21	40	19	87
The Waltons	163	10	49	36	68
Tulsa Afternoon	166	13	34	16	103
Andy Griffith	197	51	73	16	57
Scooby Doo	132	—	—	2	43
Mister Rogers' Neighborhood	—	—	—	2	10
Tulsa Christian Television	—	—	12	4	—
8:00 P.M. Average: Television					
Basic Average	225	84	92	16	33
Facts of Life	224	78	96	15	35
Movie	238	87	94	16	4
Movie	120	35	29	8	3
Movie	67	20	17	6	1
Mark Russel Comedy Special	—	4	3	—	—
Movie (IT, pay TV)	—	2	—	—	—
9:00 P.M. Average: Television					
Basic Average	188	83	84	11	10
Quincy	184	83	67	16	7
Movie	193	81	93	11	8
Movie	193	76	95	10	12
Movie	87	34	15	8	5
Special: Its Your Move	12	9	2	3	—
Movie	5	3	1	—	—

Figures in the table show that makeup of the audience will change in considerable degree according to the type of program offered and type of listener or viewer for whom appeals are provided, with an individual program. Consequently, while it is quite safe to start with the average for audience composition at the hour of broadcast, figures for the number of men, women, teenagers, and children per 100 listening or viewing homes must be adjusted upward or downward for each program, according to type of appeals provided by that program.

Whatever figures are arrived at, they are nothing more than an approximation. But the approximations can be based upon reasonably intelligent guesses, taking into account the nature of the available audience at the time of broadcast and the type of appeals provided in the program under consideration.

Effects of Program Appeals

Earlier, the proportions of men, women, and children in each of the sex-age groups that are available at each hour on weekdays were given in terms of numbers of men, women, and children available per 100 homes, assuming there are 3.3 people per home.

	Morning Hours Starting:						Afternoon Hours Starting:						
	6	7	8	9	10	11	12	1	2	3	4	5	6
Children to 13	13	47	54	35	27	29	45	23	19	35	53	62	69
Men 14–25	5	10	4	2	3	2	4	3	3	4	8	13	16
Women 14–25	10	17	13	11	10	10	10	10	10	11	17	20	21
Men 26–41	12	12	5	3	4	4	5	4	5	4	7	15	22
Men 41–55	9	10	6	4	3	3	4	3	3	3	6	13	19
Men over 56	9	11	9	8	8	8	9	8	8	8	10	13	17
Women 26–41	14	26	28	28	27	27	28	26	24	25	29	31	32
Women 41–55	12	21	21	20	20	20	19	16	16	16	19	23	25
Women over 56	11	19	22	22	22	23	21	21	22	22	22	25	27

	Evening Hours Starting:				
	7	8	9	10	11
Children to 13	70	55	27	12	7
Men 14–25	15	15	15	13	8
Women 14–25	20	20	20	19	12

Men 26–41	21	21	23	22	14
Men 41–55	20	20	20	18	12
Men over 56	20	20	21	16	8
Women 26–40	30	29	31	32	18
Women 41–55	26	26	26	23	14
Women over 56	27	27	27	22	15

Now, assuming that a given program has appeals that are strong for every type of listener or viewer available at the hour of broadcast, the makeup of the audience, in each 100 homes, tuned to that program would be approximately that shown for the hour of broadcast. No program on radio or television will have exactly the same strength for listeners or viewers in every sex-age group. Some are much more attractive to children than adults, while others are more attractive to teenagers and young adults, rather than older male or female listeners and viewers. There is always a substantial variation in attractiveness of any program for men and women.

The important question is, "Which member of the family group, taking into consideration appeals of differing strength for each person in the family, will select the program?"

Consider several possibilities. First, assume a program has very strong appeals for young children, and if tuned in, it will be dominated by young children. But there aren't young children in every home, so the program will be only tuned in within those homes where there are only young children.

Seventy percent of all children under ten years of age are found in homes where the housewife is from twenty-six to forty years of age. Consequently, adults who are included in the audience for such a program, are mostly men and women between twenty-six and forty years of age. If the program is a type most attractive for women from forty-one through fifty-five years of age, the audience includes not only women in that age group, but men of equivalent age, plus children ten to thirteen years of age, and teenagers. In a household where the housewife is from forty-one through fifty-five, there will be substantial numbers of boys and girls in those age categories.

This is typical of the American family. As a consequence, in estimating audience makeup, begin with proportions of listeners or viewers in various sex-age groups available per 100 homes. Then, it is necessary to determine the type of listener or viewer most likely to be the program selector in homes where this particular program is tuned in.

Family composition will vary in different homes, according to the age of the housewife. A final estimate must be made, based upon proportions of listeners or viewers in various sex-age groups likely to be found in those homes, including a selector of a given sex and age who tunes to that program in those homes. It is still a matter of guesswork, but these factors must be taken into consideration in making guesses as to the composition of a program audience.

Chapter 19

Attention by Audiences to Programs Broadcast

Commercial rating services show that a given radio, television, or cable program has been tuned in by a given percentage of homes in a community. Sometimes they give information concerning numbers of men, women, and children who tuned to that program for each 100 homes where the program was received. Rating services do not, and cannot, produce exact compositions of the audience for each separate home. Exact types included in the family group will vary from one family to another. Within any one selected home, there are men, women, boys, and girls representing a wide variety of sex and age groups.

In home A, for example, available at the hour the program is aired are: a man and woman, both in the twenty-six to forty age group, a fifteen-year-old girl, a 6-year-old boy, and a sixty-five-year-old woman, the grandmother of the two children who are tuning in to a program. In home B at the same hour are: a man in his late fifties, consequently, in that fifty-six to seventy age group, his wife, in the forty-one to fifty-five group, a twenty-two-year-old man, attending college, along with a sixteen-year-old sister. In other households, members of the family who are available during evening hours will show equal variation with respect to sex and age.

Listener and viewer tastes will also vary widely. Men have noticeably different preferences with respect to programs than women of the same age. Young women have different program tastes than those in older age groups. Teenagers are interested in entirely different types of programs than those strongly attracting younger children. Representing a wide variety of sex-age groups, no two members of a given family will appreciate any given program equally well. Some member of the family will enjoy a program tuned in considerably more than other programs available at the same hour of the same day.

If this is not the case, the program will not be selected in that home. But other members of the family do not share completely similar feelings for that program which characterize the program selector. Despite whether or not other family members like the program well, casually, or not at all, while it is on the air, and unless they make a conscious effort to shut themselves up in a room where it cannot be received, they will be included in the total audience for that program. In view of wide sex and age variations in types of listeners or viewers included in any one household, it follows that there will be a wide variation in degree of like to dislike for any program by different members of each family.

VARIATIONS IN ATTENTION GIVEN TO PROGRAMS

While different family members are living in a home where a given program is tuned in, there will still be a wide variation in amounts of attention given to that program. One or several family members who like the program will give it close attention, unless other activities make such attention impossible. Other family members, meanwhile, who do not like the program, will give lesser amounts of attention. Some members of the family group included in the total audience, home and within range of the set while the program is on the air, will actively dislike the program and give it very little attention. These family members make a psychological effort not to tune in at all.

The variations result from degrees of programs liked and disliked by different members of the family group. No two members of any given family will like any given program in an equal degree. Consequently, no two members will give the program exactly the same amount of attention. Another major factor affecting amounts of attention given to a program while the program is on the air concerns family members engaged in activities other than viewing, which makes full program attention impossible. There are certain necessary duties to be performed, such as preparing meals, washing clothes, reading, sewing, mending garments, caring for children, and other varieties of household duties. There are multitudes of tasks, and some of the most significant include writing letters, making telephone calls, and doing desk-work.

If the program is well liked, there will be an adjustment for the time household duties are performed, to permit giving full attention to that program. The hour the program is on the air will be kept clear for viewing by that family member who likes the program. Sometimes household tasks cannot be put off or guests drop in while that program is in progress. In those circumstances, full attention cannot be given to the program, no matter how much one may enjoy the program. Therefore, two conclusions are inevitable:

1) The degree of attention given to any program will vary appreciably for different members of the family group, in relation to the extent different individuals like the program.
2) The amount of attention given will be affected by the need for carrying on household duties. Many duties must be performed while the program is on the air.

As a result, not all program viewers, even during prime time, will give close attention to that particular program.

LEVELS OF ATTENTION

For convenience, various degrees of attention given to a program by individual family members may be classified under the following headings:

Type-5 Attention Level: The viewer likes the program very much, and is doing nothing else but watching the program, or listening in case of radio. The listener or viewer is making an effort to catch every word. This type of attention can be described as that of a small boy for a thriller Western, cartoon, or space-adventure series, such as "Black Star." For an adult, it includes attention given to a broadcast news item about a subject that directly affects that person. For adults, very few programs result in type-5 levels throughout the program. Such attention may be given to some portions, with relaxed attention to less exciting portions.

Type-4 Attention Level: This is a considerably less intense attention level than that classed as Type 5. No effort is made by the listener or viewer to follow every word closely, or every idea presented.

This is the type of attention given to a musical number, for example, or less exciting portions of dramatizations. It implies the listener or viewer is doing nothing else in addition to listening or watching.

Type-3 Attention Level: This level highlights casual attention; however, the person is doing something else while watching or listening. The greater amount of attention will be given to the program. A housewife, for example, may listen to a well-liked disc jockey music program while knitting, ironing, or conducting activities more or less mechanical.

Type-2 Attention Level: This type of attention is given to a program while the listener or viewer is engaged in some other activity while listening or viewing. The greater part of the attention level is devoted to that other activity. It is no more than "fringe" listening or viewing. The program is not disliked, or the set would be turned off. It can be characterized by students studying or housewives preparing meals while a radio or television set is turned on, providing background listening or viewing.

Type-1 Attention Level: This level characterizes those who can't help themselves. They do not like the program and would prefer the set be turned off or tuned to a different program. But someone else in the home has turned on the program, so the type-1 level is listening or viewing against a person's will, while that person makes every effort not to listen or watch. The person is doing something besides listening or viewing. By giving most of his or her attention to that other activity, the activity is not routine, because it involves concentration or thought.

These several levels of attention assume the listener or viewer is giving some attention to the program, willingly or not. If no attention whatever is given to the program, the person is sufficiently far enough away from the set and not able to hear or see the program. In most cases, any member of the family group at home, awake, and available when the program is turned on by someone in the family, on one of the sets, will be giving one of the five types of attention to that program.

VARIATIONS IN ATTENTION LEVELS WITHIN A PROGRAM

The amount of attention will change frequently. For exciting parts of a newscast especially interesting, or comedy protions of a variety program, a person may give a type-5 level of attention. It is impossible to continue this high level of attention for all elements in a sixty-minute or thirty-minute program. There will be low spots where attention may drop to the type-4 attention level. Some other member of the family may ask a question or start conversation; therefore, attention may drop to a type-3 level while the conversation is in process.

On the other hand, a person giving only a low type-2 level for a television program in a room where he cannot watch the screen may develop a higher level of attention when something is said or heard that arouses interest. They may stop what they are doing and, for a minute or two, go into the room with the television set and give a higher attention level to the program for a short period of time. There is a considerable amount of fluctuation in the amount of attention given by any person to any program. The average level for a specific listener or viewer may be relatively high or low.

How much attention is given? Amounts of high, divided, or low attention given to programs varies with no uniform method, style, or situation. In Tulsa, Oklahoma, 80 percent of all women reached during the daytime in a coincidental telephone survey reported they were listening to radio when the telephone rang and were engaged in some other activity while listening. 70 percent of those watching television at night were similarly not above the type-3 attention level. It makes a difference which member of the family group tunes in a program. Considering only prime time evening programs, women reported close attention to two thirds of those programs selected by the woman herself. Equally close attention was given to only one third of the programs selected by other members of the family group.

GENERAL CONCLUSIONS

It seems reasonable to conclude that the amount of attention given programs by various individual members of the family group varies, and also that there is variance in proportion to the extent each member of the household likes the program in question. If appeals are strong for any given member of the family group, attention will be high. If appeals provided by the program are weak, attention will normally be low, as well. The mere fact a given individual is included in the audience of a program is no guarantee that the individual will give close attention to that program.

Chapter 20

Advertising and Purpose Materials in Broadcast and Cable Programs

A radio, television, or cable program is broadcast to accomplish some specific purpose, and it should reach as large a number of listeners or viewers as needed for that purpose. Unless the purpose is simply to build an audience for the station or serve as "filler" for an unplanned open space in the station's schedule, reaching a large number of required types of listeners or viewers is not enough. The program must influence those listeners or viewers reached in some desired direction. It must sell those listeners or viewers on the desirability of buying products, goods, services; or otherwise modify attitudes; or create entirely new ones in some desired area; or convey information. Whatever the materials included in the program intended to influence attitudes, behavior, or convey information to an audience, they must be included specifically within a program schedule that accomplishes the program's purpose. This is referred to as advertising purpose materials.

In case of commercially sponsored programs, the purpose is to sell a particular kind of merchandise, or create a favorable attitude on the part of listeners or viewers for a given produce or industrial concern. It follows, therefore, that purpose materials can be the commercial announcements, or "commercials," provided within the program.

By FCC rules and regulations, these commercial announcements are clearly identified as separate elements—not part of the program as such, but included within it. The announcements may be intended to sell some specific product or create goodwill for one particular company.

Separate announcements are included in sustaining programs to influence attitudes with respect to an organization, charity, or a similar group. With a public service objective, these announcements include a variety of separate service announcements supporting that objective.

In many sustaining programs, materials are provided to accomplish the purpose of these programs. The purpose of some sustaining programs is to provide information, as is the case with certain types of educational programs. In those situations, the entire program will consist of talk or an interview with a political candidate. The purpose is to affect attitudes in directions other than creating a desire for a particular produce; this type of program will consist entirely of purpose materials.

REQUIREMENTS FOR EFFECTIVENESS OF PURPOSE MATERIALS

Whether purpose materials consist of separate announcements inserted in the program or spread throughout the entire program, requirements for effectiveness are the same. Requirements for effective purpose materials are outlined in terms of conventional commercial or public service announcements. Where special comments, relating to entire program purpose materials are appropriate, necessary comments will be added. Where purpose materials are presented as separate announcements, all of the following requirements must be met for the materials to be effective.

Purpose Materials Should Be Presented in a Favorable Climate

The entertainment portion of the program puts the listener or viewer in a receptive mood or frame of mind before the purpose materials are introduced. To ask a favor of an acquaintance, one would prefer to ask that favor at a time when he is relaxed, comfortable, and in a pleasant mood. Similarly, to attempt to sell a listener or viewer by means of commercial announcement, one has a greater chance of success if the listener or viewer is relaxed and enjoying the entertainment provided by that program.

Years ago, the Schwerin Research Company suggested, on the basis of tests for listener and viewer response to commercials, that most commercials are much more effective when presented in a program of light comedy than one with strong danger-conflict elements, or those with suspense formats. In the case of the entire program's type of purpose materials, the selling job of ideas accomplished by the program will be stronger if that program succeeds in creating a pleasant, relaxed attitude on the part of listeners or viewers.

Purpose Materials Must Harmonize with the General Mood of the Program

The requirement of unity in the structure of a program applies to purpose materials as well as other parts of the program. Purpose materials must fit the mood; otherwise, they annoy the listener or viewer and create resentment. For example, a program of easy-listening music shouldn't be interrupted with a loud, raucous, "hard-sell" commercial announcement, any more than the same program should suddenly present a "hot-rock" recording.

This was a notable and most attractive feature of the old "Johnny Martin Show," on KRMG Radio Station, Tulsa, Oklahoma, where the middle advertising purpose material commercials were presented in such a way as to be part of the entertainment function of the program. This is called "hitting them from the floor" live, rather than recording commercials on tape, and playing the commercial purpose material back over the air.

Similarly, in the old "Martin Kane, Private Eye" program, sponsored by a cigarette company, the middle commercial purpose material was integrated into the program proper. Near the middle of the program, the featured personality, a detective, always dropped into a tobacco store, where the proprietor of the store was found selling a package of the advertised cigarettes to a buyer. In the process, the owner extolled the superior qualities of that brand. The integrated commercial purpose material is rare in today's television and cable programs. In many programs, however, including most of the Quinn-Martin programs (sponsored by the Ford Motor Company), the Ernie Ford musical variety programs, and the "Dinah Shore Show" for Chevrolet, care was taken to have the featured personality lead in the commerical or provide a distinctive visual signature for the move into a break between the dramatic material in the program proper and the beginning of the commercial. This is not a necessity, but it is essential that commercials be selectively placed, so as not to break into the program's mood. If the entire program is made up of purpose materials, there is no problem.

Purpose Materials Must Capture Attention and Hold That Attention

The reason a sponsor pays for a program is to get a chance at the listener or viewer with his commercial message. Presumably, the program entertains the listener or viewer and secures their attention. If commercials, inserted in the program, are not interesting and as attention-worthy as the program proper, the listener or viewer will not be as attentive to that portion of the broadcast most essential, from the standpoint of advertising messages carried on the program.

The Schwerin Company, after conducting tests of listener and viewer responses to thousands of television commercials years ago, recommended commercials start with an attention-getting device. This is true today, because the listener or viewer must immediately be given some reason not to switch his attention away from the television or radio set whenever a commercial begins. Schwerin suggested the use of animated cartoon material to begin a commercial for television. It should entertain and capture attention. Perhaps Scherwin's material is now outdated, but certainly not the means, intent, or method. Whatever device is used, the commercial announcement should capture attention and, secondly, be no less interesting to the desired types of listeners or viewers than the program materials surrounding it.

Actually, listeners and viewers do not want to listen to or watch commercials. No one wants to be told what to think or do. Everyone has been exposed to very cumbersome commercials. As soon as they start, the tendency is to resist them and stop giving attention. To overcome this tendency, the commercial announcement must compel listener and viewer attention. It must be generally interesting and attractive enough that people give a relatively high level of attention to the entire message. All too often, sponsors become egocentric toward purpose materials. They believe they are the best announcers and television personalities—in most cases, listeners and viewers most heartily disagree!

The sponsor is wasting a great deal of money in providing a program to insure some attention for commercial announcements, when they, themselves, systematically destroy the level of attention. With respect to capturing attention, the loud fanfare of trumpets method, use of a "shocker," and fear-creating material are by no means the only, or best methods of securing attention of listeners or viewers. The commercial can be entertaining, from its beginning, in its own right. Those middle commercials in Monday Nite Football, featuring Millers' Lite Beer, with a host of sports personalities, are frequently high in entertainment value, and receive just as much attention, as any other element in the program.

Purpose Materials Should Give the Listener or Viewer Some Reason for Accepting, and Acting Upon, the Sponsor's Message

Sometimes, the reason is stated directly in so many words. We can save money by shopping at Brown's store. We can be assured of high quality if we buy "Gold Spot" products.

These are actual cases: "Dow: There's good chemistry between us"; "You can be sure if it's Westinghouse"; "G.E.! We bring good things to life!" In other cases, the reason given is implied, rather than stated outright. One company suggests that its brand of cologne is appropriate to virile, masculine types, by using actors who represent the rugged cowboy or horseman. Other role models include athletes, businessmen, deep-sea divers, and car-care mechanics, using and testifying to the products, goods, or services advertised.

In the course of one advertising campaign, a soap manufacturer suggested that women who used the advertised brand of soap were unusually attractive to those of the opposite sex. By having the product endorsed by debutantes, and by following that endorsement with a statement that the girl featured in the commercial was actually engaged to be married, the advertiser suggests one reason to accept and act upon his message. With women's sports today, many female athletes endorse products that are acted upon, and accepted, by listeners and viewers. Some commercials give reasons for accepting the product or idea. For a program whose purpose is to sell an idea rather than a product, the reason given does not have to be a logical one. Ivory soap, for years, was advertised on the basis of one distinctive quality: "It floats!"

Anacin for years gave listeners and viewers no better reason for using the product than the statement, "Like a doctor's prescription! It is made up of a variety of ingredients." Lucky Strike cigarettes, before cigarette advertising was banned on radio, television, and cable by the Federal Communications Commission and the Federal Trade Commission, were advertised during one major campaign by the use of the slogan, "Be happy, go Lucky!" And for a longer period, the same cigarette offered as a feature to its advertising that the tobacco in the cigarette was "toasted."

Ford cars told you that "You're ahead with a Ford, all the way," without bothering to tell you why. Chevrolet made use of the singing slogan, "Baseball, hot dogs, apple pie, and Chevrolet!" If something pleasant can be associated or identified with the product—the idea of a vacation in pleasant surroundings, the picture of a man well fed and relaxed after having a substantial meal, the picture of a happy, contented baby—the pleasant feeling carries over and identifies with that product.

When the listener or viewer thinks of that particular product, he thinks of it with a pleasant feeling. Not all elements, however, relate to giving the listener or viewer a reason to believe or accept the program, of which purpose material is a major part. Purpose materials, therefore, must be used in good taste. For example, the use of an oven cleaner product, preceded by portions of the extermination of Jews in Nazi German death camps in the program series "Holocaust" is a classic example. Use of this product gave listeners and viewers an extremely negative reaction.

Purpose Materials Should Have Substantial Emotional Values

There is an obvious reason for this recommendation. Use of emotional stimulation produces a heightened feeling on the part of listeners or viewers, as well as greater listener/viewer response to the message of the commercial. Consequently, by use of pictures relating to happy, healthy babies, the identification process is excellent for selling soaps. Use of five- and six-year-old children eating breakfast foods with obvious enjoyment is essential to heightened listener/viewer feelings. Families sitting down together at the breakfast or dinner table, or getting ready for a ride in the new family car, both have emotional value that others can identify.

Attractive girls who use an advertised brand of designer jeans provide emotional value. Young couples wearing these jeans, seeming to be much in love, all provide emotional value; with women wearing these jeans and having fun, there is also a natural progression to follow these emotional stepping stones through basic identification. In the public-service field, the little girl with the braces on her legs, who was a symbol of drives for funds by the March of Dimes Infantile Paralysis Organization, and the poster children for the Jerry Lewis Muscular Dystrophy Telethon, both provide identifiable symbolism for public service.

Without question, use of emotional stimulation or sex appeal helps reinforce the sponsor's message. When the entire program is composed of purpose materials, there is just as much reason to make use of materials with emotional values that make the listener or viewer feel something as there is in the case with commercial announcements.

Purpose Materials Should Include Some Element Making It Easy for the Viewer to Remember the Message, Long After Hearing or Seeing the Broadcast

Sponsors for programs want purpose materials to achieve results, not only at the moment when a person watches or hears the commercial announcement, but the next day, a week later, and even months later. The actual buying of an advertised product rarely takes place within a space of thirty or forty seconds, or even minutes after the listener or viewer is exposed to a commercial message. Sponsors want the entire commercial, or some portion of it, to be remembered, the part that will express the message essence of that produce.

A variety of methods are available for giving the commercial announcement essential qualities. Use of a slogan, "You can be sure if it's Westinghouse," or "Wonder helps build strong bodies twelve ways," or "You only go around once, so go for all the gusto you can," can be weak if the name of the product is not included. Nevertheless, listeners or viewers will not forget brand names or slogans that fit. "Lite has a third less calories than regular beer, and tastes great" is a much better qualification. Another essential quality concerns use of musical jingles. One of the earliest began with radio's "Pepsi Cola hits the spot." Others, important in radio before the development of television and cable, includes "Rinso White's little washday song" and Campbell Soup's "mmmmmmmm good . . . That's what Campbell Soups are, mmmmmmmm good!"

More recently, the highly effective "Winston tastes good, like a cigarette should" was set to music. A well-known light-beer maker used the rhythmical talking jingle such as, "Hey! This beer is for all you guys that stack 'em, wrack 'em, and sack 'em!" This included a musical twist at the end of the song and the patter jingle. Post cereal's "just a little better than any other cereal happens to be," the "Merry Oldsmobile" song used as the theme in Oldsmobile advertising, as well as the sign-off signature for the "Dinah Shore Show," "See the USA in your Chevrolet," were all part of song and patter jingles that national advertisers sought to localize in their products. And within a twenty-second musical "bed," an opportunity was given the local distributor to insert his name, address, and telephone number. They were presented as part of a flexibility effort to give a localizing aspect to major national advertisers.

Use of music as a sponsor-signature theme has been effective for years in spelling out "JELLO," while another signature idea is the special method of using visual motion pictures or stills, identified with advertising for a particular product. Kool Cigarettes' visual use of the penguin, combined with either "Change from hots to Kools" or "Smoke Kools," was a good example during the cigarette heyday on television.

The practice of Coors beer introducing all their commercials with pictures of "cool rocky mountain spring water" and "the high country," showing a lake nestled in the hills or mountains of Colorado, are excellent visual signature methods. Gillette's use of baseball player testimonials, as well as the long-continued series of cartoon commercials featuring the cartoon character "The Old Pro" and his sidekick "Elwood" were highly successful in years past.

Even program package and syndication companies have highlighted visual signatures. It is difficult to forget the syndication trademark of the late Jack Webb's mark VII Limited Production Company, with the arm and hammer indentation blotter, marking the VII symbol in stone with a full live-arm, hand-and-hammer shot. Obviously, these are not the only methods used to give purpose materials a memorable quality. Use of novelty and information related to listener or viewer interests are also effective. Talks or documentaries that intend to influence attitudes by making use of vivid illustrations while helping make points memorable are using subtle methods that enforce listeners or viewers to remember. Whatever method is used, purpose materials are most effective if listeners or viewers remember them for days or months after seeing or hearing them on the air.

Purpose Materials Should Not Create Audience Antagonism or Opposition

If the listener or viewer is sufficiently antagonized, he may recall the commercial and product advertised and deliberately refuse to act along lines desired by the advertiser. There are a number of facts about commercials that annoy listeners or viewers. When someone is annoyed when a commercial is inserted into a program without advance preparation and planning, and where that commercial intrudes on the flow of the ideas of the program, you may rest assured that commercial will not be recalled or received. Many will be annoyed when too many commercials are inserted or "stacked," in a single program, or when a single commercial seems needlessly drawn out or prolonged.

Listeners and viewers will be annoyed and irritated if the subject matter is unpleasant or falls short of the requirements of good taste. Audiences will be annoyed and resistant if the commercial is negative toward accepted ideas or if exaggerated claims are made, and the hard-sell technique is consequently overworked. In some degree, people are annoyed if claims in the commercial reflect unfavorably on the advertiser's competitors and if people are instructed to buy a product, or do something without reason, provided for such action. Listeners and viewers want to be treated like thinking adults, not children. It is certain listeners and viewers will be annoyed and will develop strong opposition to accepting the message conveyed, if a scolding attitude is used. When listeners and viewers are told outright or by implication they are lacking in intelligence; or that they fail in meeting standards of personal and social morality; or if audiences fail in meeting standards of personal and social morality; or if audiences fail to follow instructions given in the message; these are all excellent examples of annoyed opposition.

It is easy to find instances of commercial announcements creating listener or viewer opposition. Such announcements are less effective than if they encouraged acceptance. It is easy to find illustrations of this shortcoming when entire programs are intended to instruct, raise standards of appreciation, or modify existing attitudes. This is one of the very common failures of current educational programs. For example, a program whose objective is to raise cultural standards of music appreciation will fail if the featured personality states, or even suggests, that anyone who does not appreciate a certain kind of music lacks culture. That is obviously a direct slap at those listeners or viewers in the audience who don't care for that kind of music, which is being presented as "good music" on the program.

No one likes to be told he is guilty of poor judgment or low standards. Consequently, if a show commits this judgmental error, the program and purpose materials have failed. The kind of listener or viewer whose music tastes need refinement will be influenced in exactly the opposite direction from that intended on the program.

A program may be presented with the purpose of creating greater social acceptance for those of other races. If the program tells listeners or viewers that, in effect, they hold prejudices, or they are immoral or guilty of bad judgment, then those listeners or viewers who do hold such prejudices find those prejudices strengthened, rather than lessened, as a result of the program.

If a Republican political candidate, attempting to win votes of non-Republicans, attacks the integrity of all Democratic office-holders and the good judgment of all candidates, he certainly isn't likely to induce any loyal Democrats to switch over to his point of view. He will not be effective in winning votes of either Democrats or independent voters, if he devotes all his time to praising the infallible right of the GOP and all who bear the Republican label and hold office.

If the purpose materials even suggest the listener's or viewer's existing views are bad and he, consequently, is showing a lack of good sense or moral behavior in holding such views, the program will not succeed in influencing listeners or viewers to change those views that are attacked or ridiculed. Whether in an entire propaganda program, or even in the form of commercial announcements within a program, purpose materials must be so presented as not to create listener or viewer antagonisms. If they do, they will fail to accomplish their purpose.

Chapter 21

Total Requirements of an Effective Program

The greater part of this book has been devoted to consideration of programs from different points of view. The object throughout has been to provide various tests or standards on the basis of judging effectiveness of given programs. The most important single purpose of the entire book has been to give a set of standards for the evaluation of programs. This final chapter attempts to summarize various tests suggested and to provide a series of specific tests that can be applied to any program to determine that program's effectiveness. The tests provided deal with effectiveness of the program and its success in achieving the purpose it is broadcast. No attempt is made to evaluate the artistic worth of programs. Frequently, artistic quality has little or nothing to do with effectiveness. However, one program test is added that does not relate to effectiveness in achieving the program's purpose but which deals with overall effects of the program on listeners and viewers. Whether or not the total social effect of the program is a good one, or at least one that is not socially injurious, should be considered as well.

The various tests are as follows.

WHAT IS THE MAJOR PURPOSE OF THE PROGRAM?

Obviously, in attempting to evaluate a program's success in achieving its purpose, one must know for what reason the program was put on the air. The purpose relates to the person or agency which pays costs of presenting the program. This includes the sponsor, if the program is sponsored, who pays costs of presenting a sustaining program when those costs are paid by an outside agency, and costs of time donated by the station. This also includes the organization on whose behalf a network or station presents a program, when part or all of the costs are paid by the network or station. The station itself has a purpose if the program is presented on a sustaining basis to heighten network or station prestige, or build a larger audience for a single station or all stations comprising that network. This includes specific programming during network "sweep weeks," when ratings are taken.

Keep in mind that the purpose of a program is never that of entertaining listeners or viewers. Entertainment is simply a means to an end. In some cases, the basic purpose may be that of informing the audience. In these instances, information that is provided will be given either to attract listeners or viewers or to contribute to a more fitting basic purpose of influencing listeners' and viewers' attitudes.[1] A great many sponsored programs are presented with the purpose of selling or creating a greater demand for a particular product or service supplied by the sponsor. In some cases, the purpose may be to create goodwill for the sponsoring company and to enhance prestige for that company in the community or throughout the nation, in the case of a network program.

If the program is sponsored, the purpose may be to influence listeners' or viewers' attitudes on social or economic problems. And, in case of a religious program, it is to influence attitudes with respect to religion. In the case for political broadcasts, the purpose is to influence political thinking in such a way as to win votes for a single candidate or group of candidates wearing the label of one specific political party or pressure group. If the program is presented on a sustaining basis, the purpose is to influence attitudes or behavior of listeners or viewers in some manner. Attitudes are those relating to social, religious, or economic concerns. Occasionally, a station may present a program by taking an editorial position with respect to a given political issue. Sometimes, an "attitude" program may be one where the purpose is to influence listener or viewer attitudes with respect to the station, cable facility, or network, to make listeners or viewers have a more favorable attitude toward the broadcasting company. Some sustaining programs are presented primarily to build a larger audience for the station at a particular time. In most cases, where audience-building is the objective, there is also some degree for station-prestige-building in the presentation of the program.

In any case, every program broadcast is put on the air for primary purposes of justifying a sponsor's spending dollars for a local program, or thousands of dollars in network programs, to put that program on the air. If the program is broadcast on a sustaining basis, the purpose is to advance the interests of a public service organization, create prestige for the station, or build a larger listening or viewing audience for the station or cable outlet.

WHAT KIND OF LISTENERS OR VIEWERS MUST BE REACHED BY THE PROGRAM TO ACHIEVE THE DESIRED PURPOSE?

To accomplish its purpose, every program must be heard, or seen, by a specific type of listener or viewer who will be useful from the point of that program's purpose. If the purpose is to sell merchandise, the program must reach those individuals in the population who do the buying of that particular kind of merchandise. If the purpose is to influence attitudes, or those who are "neutral," hopefully they can be influenced in the desired direction. If the purpose is to win votes for a candidate, the program must reach voters, in particular those voters who have not yet made up their minds with respect to several possible candidates. If the purpose is to build prestige for the station, the program should reach those individuals whose opinions of the station, as an outstanding local station, are important. This should include opinion leaders, community leaders, and a general adult audience more receptive to efforts made in other programs to influence attitudes and behavior. This is in respect to buying of products and the observation of opinions held on various problems. If the purpose is that of building a bigger audience for the station at some given hour of the day, the program should be aimed at the type of listener or viewer, who, at that hour, is available in the greatest numbers. In particular, care should be directed at that portion of the highest available audience not being served effectively by programs presented on other stations.

The types of listeners or viewers required include a type needed most and, in addition, other types from the standpoint of sex and age that are useful from the basis of that program's purpose. In a majority of cases, identification of needed listeners or viewers, whether needed most or useful, should go further than a consideration of sex and age. The needs of listeners or viewers with buying power is indicated by economic status, existing attitudes, political preferences, and similar characteristics.

DOES THE PROGRAM REACH A LARGE ENOUGH NUMBER OF NEEDED TYPES OF LISTENERS OR VIEWERS?

Whether or not the program will reach a sufficiently large audience of needed listeners or viewers depends on several factors.

Is the Program Broadcast by a Station Attractive to Listeners or Viewers of the Type Needed?

This is not an absolute requirement. Nevertheless, a sufficiently good program, broadcast at the right hour, can attract large numbers of needed listeners or viewers, even if broadcast over a station less effective in reaching listeners or viewers required for this program. But it will require expenditure of considerably more money in promotion of the program and will take a good many more weeks to let needed listeners or viewers find out about the program, if it is carried on a less desirable station than one that is already a favorite for the type needed for this particular program. Therefore, selection of the station may be an important factor.

Is the Program Put on the Air at a Time that Permits It to Be Most Effective in Reaching Needed Types of Listeners or Viewers?

Time of broadcast is extremely important. First, the time the program is put on the air must be one that provides as many listeners or viewers as possible that are available. The best program, broadcast at 3:00 A.M., is not likely to reach as many housewives as a program broadcast at 10:00 A.M. One at 2:00 P.M. will reach a much smaller audience of men from twenty-six to forty than one broadcast at an hour when more men of that age group are at home, and available, while other things are equal.

In addition to importance to the availability of needed listeners of viewers, time of broadcast is important from the standpoint of the nature of competing stations at the same hour for both needed listeners or viewers.

How Strong Are the Total Appeals Provided by the Program, and How Attractive is the Program for Needed Listeners or Viewers, as Well as a General Audience?

Even if broadcast at the most desirable hour, a program will be relatively ineffective in attracting large numbers of needed listeners or viewers if weak appeals are provided for listeners and viewers. To attract an audience, the program should provide at least two or more of the seven basic appeals, which include conflict, comedy, sex appeal, human interest, emotional stimulation, information, and importance. It should provide a feeling of involvement for needed listeners or viewers and offer believable characters and situations, as characterized by a reasonable degree of freshness, novelty, and originality.

The appeals offered should be strong, for those types of individuals needed for purposes of the program. It should also provide as many appeals as possible for listeners or viewers who want to listen or watch the program. The appeals provided must be moderately strong for every type of listener or viewer likely to be available in the ordinary family group at the time of broadcast. In a substantial number of homes, the individual who serves as program selector at the hour of broadcast may be different from individuals who might be classed as members of the needed audience. The program must be tuned in by the selector before those in the needed audience have a chance to listen or watch. In some degree, the rating received by the program may serve as an index to that program's effectiveness in recruiting needed listeners or viewers. Certainly, if the rating is extremely low, the program cannot be captured by any substantial number of needed listeners or viewers. However, even a program with a high rating may include in its audience only a small minority of needed types. It may be broadcast at a time when few needed types are available, or its appeals may be such as to make the program attractive to types entirely different from those needed to satisfy the program's purpose.

The rating must be moderately high, at least, to permit presence of any substantial number of needed types. In addition, the appeals offered must make the program attractive enough to needed types so a large proportion of those needed types are available and will be tuning in.

DOES THE PROGRAM ATTRACT SUBSTANTIAL NUMBERS OF NEEDED LISTENERS OR VIEWERS AT A REASONABLE COST PER THOUSAND OF NEEDED LISTENERS OR VIEWERS?

We have no figures available as to the cost of programs in terms of numbers of needed types reached. We do have figures that give an approximation of the cost of each program, per thousand homes reached. Figures that give a figure for the cost of each program per commercial minute can be closely analyzed. The cost per thousand homes is reached by projecting the rating to determine the total number of homes tuned to a program during the average minute it is broadcast and dividing the total cost of time and production for the program by the number of thousands of such homes. The cost per commercial minute assumes that the program devotes industry-approved amounts of time for commercial announcements. Breaks generally occur at the seven and a half and 15-minute marks in every fifteen-minute segment; for commercials, there are six minutes for each sixty-minute program, three minutes for each thirty-minute program, and two and one half minutes for each fifteen-minute program broadcast during evening hours. This assumes a sixty-minute program; cost-per-thousand homes per commercial minute is estimated by dividing the cost-per-thousand homes reached by six, the number of commercial minutes allowed.

It isn't necessary for us to know exact costs of the program per commercial minute, per 1000 needed listeners or viewers in general, or whether the program reaches needed listeners or viewers at a reasonable cost. We can estimate whether the cost is well above average, average, or below average. This may give us a close approximation to answer the question constituting the fourth test of program effectiveness.

DOES THE PROGRAM HOLD THE ATTENTION OF THOSE NEEDED LISTENERS OR VIEWERS WHOM IT DOES REACH?

If the program is to be rated as effective, it must not only reach needed listeners and viewers, it must also hold their attention. Its ability to hold a relatively high level of their attention depends on these factors.

Is the Program Broadcast at an Hour When Most Listeners or Viewers of Needed Types Are Able to Give Attention?

One factor that influences attention is the extent to which needed listeners or viewers engage in other activities while tuned to the program. If the program comes at a time when other activities are carried on, then the audience will not be able to give close attention to the program. For example, a program broadcast during the time when housewives are busy in the kitchen preparing meals or busy in the kitchen washing dishes after a meal are not likely to receive a great amount of attention from those housewives. For certain kinds of needed listeners or viewers, the hour of broadcast may be an important factor in determining whether a program receives a high level of attention from the kinds of listeners or viewers required for accomplishment of the program's purpose.

Is the Structure of the Program Effective Enough to Make It Easy for Listeners or Viewers to Give the Program a High Level of Attention?

If the structure of a program is weak, listeners or viewers will be able to give attention only at the cost of a considerable amount of effort. They are likely to become tired; therefore, attention wanders into other fields. For complete attention, the program must satisfy every one of the seven requirements of good structure, including program unity, effective opening and closing, strong and fast start, variety in types of material used, a sufficient degree of unit-to-unit change and contrast, good pace with an effect of moving along, and effective use of building and climax. These requirements were discussed in detail previously in this book.

Does the Program Offer Strong Appeals for the Types of Listeners or Viewers Needed?

Use of appeals has been discussed in relation to recruiting sufficient numbers of needed listeners and viewers. But there is also a definite relationship between the appeals used and amount of attention a program receives. As noted previously, attention is closely related to the extent a person likes a program. Likes and dislikes on the part of any person depend on the extent the program provides those appeals that are strong for listeners or viewers of that specific type. Consequently, a major test of the amount of attention a program receives from needed listeners or viewers is the extent the program provides appeals that are strong enough for those needed listeners or viewers.

In addition, does the program give those listeners or viewers the required feeling of being involved in the program?

If the level of attention given by needed listeners or viewers is to be high enough to allow the program to be effective, the program must meet all three of the tests: broadcast at an hour when needed people can give undivided attention, provide effective structure, and offer appeals that have high values for particular kinds of listeners or viewers necessary for the program to reach.

IN A FINAL TEST OF EFFECTIVENESS, DOES THE PROGRAM PROVIDE PURPOSE MATERIALS THAT ARE EFFECTIVE IN INFLUENCING NEEDED LISTENERS OR VIEWERS?

A first requirement of effectiveness in achieving the purpose of the program is attracting a sufficiently large number of people of the needed type and attracting a sufficiently large number of people of the needed type and attracting them at a reasonable cost. Second, having induced these people to tune in, it is important to insure that they give the program a high level of attention. But having the listeners or viewers, and having them give attention, means nothing if the purpose materials are not effective in influencing those needed people in the desired direction.

To be effective, the purpose materials must be presented in a favorable climate. They must harmonize with the mood of the program. Also, they must arouse listener or viewer interest by being interesting and pleasing. Thirdly, they must give the listener or viewer reason, not necessarily logical, for accepting the ideas presented. They should present the message in language and copy that the audience will remember. They are stronger if emotional values are provided and must not be so presented as to create antagonism, opposition, and unwillingness to accept the idea.

The test of effectiveness of purpose materials is no less important than attracting needed types of people and inducing them to give a high level of attention to the program.

SOCIAL CONSIDERATIONS

Tests of program effectiveness have been considered and evaluated on the basis of their probable success in accomplishing the purpose for which they are broadcast. This is not the only basis on which a program could, or should, be evaluated. Programs influence listeners and viewers. From a social point of view, the kind of effect a program may have on its listeners and viewers becomes important. One basis of program evaluation frequently used by professional critics is the extent the program meets certain artistic standards. This relates to social effects, and the presentation of programs that have outstanding artistic literary and cultural values should contribute to raising artistic and cultural standards on the part of the total public. However, no attempt is made in this book to erect standards of artistic appreciation of programs. That is a field entirely distinct from that dealt within this book. However, in evaluating any program from the standpoint of ultimate effectiveness, attention must be given to social considerations. These can perhaps be best considered in the form of answers to two question.

Does the Program Have a Socially Desirable Basic Purpose?

If the purpose of the program is harmful, the program cannot be classed as a good program, even if it fully accomplishes that purpose. Socially desirable purposes, in our society, are not limited to educating the public, raising cultural standards, or making our society better than it would otherwise be. In our society, advertising is considered socially acceptable. The selling of goods or services, if they are of good quality, is similarly acceptable. The fact that a program sells merchandise, or propagandizes in behalf of a certain point of view, does not, in itself, make that program socially desirable.

However, that program may be socially undesirable, or injurious, if goods sold are of a type that would injure the consumer, or if sold by means of false, misleading, fraudulent claims, or if the program with a propaganda purpose attempts to influence listeners or viewers in the direction of socially harmful attitudes or beliefs.

Does the Program Unintentionally Exert an Influence or Create Beliefs that Might Be Harmful?

From a sociological point of view, there is greater danger of this shortcoming being found in these programs than those planned intentionally to harm someone. A few questions will point up the problems. Is a program socially desirable if it directly, or by implication, questions our form of government or existing moral standards? Is it socially desirable if it encourages children not to respect parents, or disobey instructions given by parents? Is it socially desirable if, by implication, it upholds disrespect for law and order, by portraying famous criminals as desirable characters? Is it socially desirable, if it, by implication, suggests moral codes to sexual relations and alcohol, as well as drugs, that are obsolete and should be ignored?

Such suggestions have been found in broadcast programs. In the vast majority of cases, this is without broadcasters being aware of the direction the material in the program would be interpreted by listeners and viewers, especially those immature listeners or viewers. In making a total evaluation, attention cannot be concentrated wholly on the program's effectiveness in achieving a particular purpose. Some attention must be given to the total effects, unintended or intended, the program may have on listeners or viewers.

NOTES

1. Please note the bibliography on attitudes and attitude change. It also has been suggested that attitude research and theory have the potential for clarifying some of the critical conceptual and methodological problems researchers continue to run up against in the important area of personality. Some of the theoretical and empirical issues of

affective-cognitive relatins that are common to both personality and attitude concepts can be most enlightening. In the included bibliography, which goes beyond the scope of this book, there are some developments in contemporary attitude research that I consider to hold promise for our efforts to evaluate the theoretical maturity in the attitude-personality concept and perspective.

Also in this regard, the book *Motivation and Personality*, by Abraham Maslow, is excellent source material. Applying this to the broadcast and cable program, with actors or speakers, there are a series of needs and goals that can be met by demonstrating acceptance of those aforementioned recommendations.

a) physiological needs
b) safety needs
c) belongingness and love needs
d) esteem needs
e) self-actualization needs

Bibliography

Atkinson, R.; Carterette, R.; Kinchla, R. "The Effect of Information Feedback upon Phychological Judgments." *Psychonomic Science* I (1964).

Atkins, L., Bieri, J. "Ego Involvement of Anchoring of Social Stimuli." *Journal of Abnormal Social Psychology* 52 (1965).

Atkins, L.; Kujala, K.; Bieri, J. "Latitude of Acceptance and Attitude Change." *Journal of Abnormal Social Psychology* 56 (1965).

Berkowitz, Larry, "The Judgmental Process in Personality Functioning." *Psychology Review* 67 (1960).

Bieri, J.; Atkins, A.L.; Briar, S.; Leaman, K.; Miller, L.; Tricopci. *Clinical and Social Judgment: The Discrimination of Behaviorial Information.* New York: Wiley, 1966.

Bieri, J., "Cognitive Structures in Personality." In *Information Processing: A New Perspective in Personality Theory*, edited by H.M.Schroeder and Paul Suedfield. New York: Ronald, 1969.

Bonarious, J.C.J., "Research in the Personal Construct Theory of George A. Kelley: Role Construct Repertory Test and Basic Theory." In *Progress in Experimental Personality*, edited by B.A. Maher. New York: Academic, 1965.

Easterbrook, J.A., "The Effect of Emotion on Cue Utilization and the Organization of Behavior." *Psychology Review* 66 (1959).

Eriksen, Charles W., Wechsler, H., "Some Effects of Experimentally Induced Anxiety upon Discrimination Behavior." *Journal of Abnormal Social Psychology* 51 (1955).

Galenter, E., "Contemporary Psychophysics." In *New Dimensions in Psychology*, edited by R. Brown, E. Galenter, W.H. Wess, G. Mandler. New York: Holt, Rinehart, and Winston, 1962.

Gutman, N., Kalish, H.I., "Discriminability and Stimulus Generalization." In *Psycology: A Study of a Science*, edited by S. Koch. New York: McGraw-Hill, 1956.

Gutman, H., "Laws of Behavior and Facts of Perception." In *Psychology: A Study of a Science*, edited by S. Koch. New York: McGraw-Hill 1956.

Hinckley, E.D., "The Influence of Individual Opinion on Construction of an Attitude Scale." *Journal of Abnormal Social Psychology* 3 (1932).

Hinckley, E.D. "A Follow-up on the Influence of Individual Opinion on the Construction of an Attitude Scale." *Journal of Abnormal Social Psychology* 67 (1963).

Johnson, D.M., King, C.R. "Systematic Study of End Anchoring and Central Tendency of Judgment." *Journal of Experimental Psychology* 67 (1964).

Katz, D., Scotland, E. "A Preliminary Statement to a theory of Attitude Structure and Change." In *Psychology: A Study of a Science*, edited by S. Koch. New York: McGraw-Hill, 1959.

Kelley, George A. *The Psychology of Personal Constructs*. New York: Norton, 1955.

Klein, George S. "Cognitive Control and Motivation." In *Assessment of Human Motives*. edited by G. Lindzey. New York: Holt, Rhinehart and Winston, 1958.

Kujala, K., Bieri, J. "Latitude of Acceptance in Judgments of Masculinity-Femininity." *Journal of Abnormal Social Psychology* 56 (1965).

Marlowe, David, Gergen, Kenneth J. "Personality and Social Interaction." *Handbook of Social Psychology*, edited by G. Lindzey. Massachusetts: Addison-Wesley, 1969.

Miller, H., Bieri, J. "End Anchor Effects in the Discriminability of Physical and Social Stimuli." *Psychonomic Sciences* 3 (1965).

Rappaport. "The Structure of Psychoanalytic Theory: A Systematizing Attempt." In *Psychology: A Study of a Science*, edited by S. Koch. New York: McGraw-Hill, 1959.

Sherif, C.W.; Sherif, M.; Nebergall, E. *Attitude and Attitude Change: The Social Judgment Involvement Approach*. Philadelphia: Saunders, 1965.

Sherif, M., Hovland, Carl. *Social Judgment: Assimilation and Contrast Effects in Communication and Attitude Change*. New York: Yale Press, 1961.

Storms, L.H., Broen, W.E. "Drive Theories and Stimulus Generalization." *Psychology Review* 73 (1966).

Ward, C.D., "Ego Involvement and the Absolute Judgment of Attitude Statements." *Journal of Abnormal Social Psychology* (1965).

Zavalloni, M., Cook, S.W. "Influence of Judges' Attitudes on Ratings of Favorableness of Statements About a Social Group." *Journal of Personality and Social Psychology* 1 (1965).

DEAN COLLEGE LIBRARY

3 0423 00030 6603

63493
HE   1500
8700.7
.C6   McSpadden, M.R.
M175   Broadcasting and
       cable

DATE DUE

DEAN JUNIOR COLLEGE
E.R. ANDERSON LIBRARY

99 Main Street

Franklin, MA 02038-1994

DEMCO